Social and Cultural Studies of Robots and AI

Series Editors
Kathleen Richardson, Faculty of Computing, Engineering, and Media,
De Montfort University, Leicester, UK
Teresa Heffernan, Department of English, St. Mary's University,
Halifax, NS, Canada

This is a groundbreaking series that investigates the ways in which the "robot revolution" is shifting our understanding of what it means to be human. With robots filling a variety of roles in society—from soldiers to loving companions—we can see that the second machine age is already here. This raises questions about the future of labor, war, our environment, and even human-to-human relationships.

Ann Oravec
University of Wisconsin–Whitewater
Whitewater, WI, USA

ISSN 2523-8523　　　　　　　　ISSN 2523-8531　(electronic)
Social and Cultural Studies of Robots and AI
ISBN 978-3-031-14012-9　　　ISBN 978-3-031-14013-6　(eBook)
https://doi.org/10.1007/978-3-031-14013-6

© The Editor(s) (if applicable) and The Author(s), under exclusive license to Springer Nature Switzerland AG 2022

This work is subject to copyright. All rights are solely and exclusively licensed by the Publisher, whether the whole or part of the material is concerned, specifically the rights of translation, reprinting, reuse of illustrations, recitation, broadcasting, reproduction on microfilms or in any other physical way, and transmission or information storage and retrieval, electronic adaptation, computer software, or by similar or dissimilar methodology now known or hereafter developed.

The use of general descriptive names, registered names, trademarks, service marks, etc. in this publication does not imply, even in the absence of a specific statement, that such names are exempt from the relevant protective laws and regulations and therefore free for general use.

The publisher, the authors, and the editors are safe to assume that the advice and information in this book are believed to be true and accurate at the date of publication. Neither the publisher nor the authors or the editors give a warranty, expressed or implied, with respect to the material contained herein or for any errors or omissions that may have been made. The publisher remains neutral with regard to jurisdictional claims in published maps and institutional affiliations.

Cover illustration: gremlin/Getty Images

This Palgrave Macmillan imprint is published by the registered company Springer Nature Switzerland AG
The registered company address is: Gewerbestrasse 11, 6330 Cham, Switzerland

Jo Ann Oravec

Good Robot, B Robot

Dark and Creepy Sides of Robotics, Aut Vehicles, and AI

palgrave
macmillan

This book is dedicated to my late mother, Marion Oravec, whose amazing managerial prowess and consulting talent enhanced countless situations as well as rescued my career many times. It also is dedicated to the work of Professor Robert Miller, who in 1983 at the University of Wisconsin at Madison set a pioneering direction for how robot developers, anthropologists, and implementers would proceed in their efforts. My life partner, Dr. Robert Reuschlein (founder of the Real Economy Institute and Nobel Peace Prize nominee), also contributed inspiration and insights in the long journey from first page to finished manuscript. I would like to dedicate the book as well to Ralph Nader for his lifetime of public service and consumer advocacy. His 1965 book Unsafe at Any Speed: The Designed-in Dangers of the American Automobile *focused attention on technological mishaps and stimulated designers to rethink their approaches. I attended several of his workshops, which empowered me to be proactive in tackling large sociotechnical issues as well as human rights challenges.*

The book is also dedicated to the memories of Wanda Holbrook, Robert Williams, Umesh Ramesh Dhake, Elaine Herzberg, and so many others killed and injured by robots and autonomous vehicles.

Acknowledgements

I would especially like to thank everyone at Palgrave Macmillan who was involved with this book, with Rachael Ballard and Shreenidhi Natarajan putting in extraordinary effort. Special thanks to Professors Kathleen Richardson, Cathrine Hasse, and Teresa Heffernan for their pioneering research and activism initiatives as well as their leadership in the Social and Cultural Studies of Robots and AI series, of which this book is a part.

My affiliation with the Robert F. and Jean E. Holtz Center for Science & Technology Studies, University of Wisconsin at Madison, has always been a source of joy and inspiration.

Thanks also go to my colleagues in the College of Business and Economics at the University of Wisconsin at Whitewater in the Department of Information Technology and Supply Chain Management, especially for the efforts of John Chenoweth, Paul Ambrose, and Andrew Ciganek, as well as the brilliant mathematics and history of science professor Thomas Drucker. My work with the International Education for Sustainable Development Alliance (INTESDA) has been stimulating, with the efforts of Tim Desmond and Michael Sasaoka as wonderfully motivating and supportive. Professor Richard Kyte of Viterbo University has generated (and shared) many leadership and ethics insights, and Dr. Thomas Kaczmarek of Marquette University has provided creative directions in cybersecurity studies. I'd especially like to thank my students both at UW-Whitewater and UW-Madison for their brilliant ideas and insights.

Any errors or omissions in this document remain clearly my own, along with the various flights-of-fancy, delusional rantings, and dark meanderings.

Contents

1 Entering the Second Century of Robotics and Intelligent Technologies: An Opening Note — 1

2 Dramaturgical and Ethical Approaches to the Dark Side: An Introduction — 11

3 Negative Dimensions of Human-Robot and Human-AI Interactions: Frightening Legacies, Emerging Dysfunctions, and Creepiness — 39

4 Love, Sex, and Robots: Technological Shaping of Intimate Relationships — 91

5 The Long Robotic Arm of the Law: Emerging Police, Military, Militia, Security, and Other Compulsory Robots — 125

6 Gilding Artificial Lilies: Artificial Intelligence's Legacies of Technological Overstatement, Embellishment, and Hyperbole — 153

7 "Our Hearts Go Out to the Victim's Family": Death by Robot and Autonomous Vehicle — 177

8 Robo-Rage Against the Machine: Abuse, Sabotage, and Bullying of Robots and Autonomous Vehicles — 205

9 The Future of Embodied AI: Containing and Mitigating the Dark and Creepy Sides of Robotics, Autonomous Vehicles, and AI   245

Index   277

CHAPTER 1

Entering the Second Century of Robotics and Intelligent Technologies: An Opening Note

As the cases and examples in this book *Good Robot, Bad Robot* demonstrate, the problem of "bad robots" may not be solved by making robots seem more human. We may be living with robots, automated vehicles, and other AI-related entities that many of us perceive to be "dark" and "creepy" for many years to come. Some of these dark traits are the result of designers' decisions, such as the manners in which certain robots apparently elicit fear on the part of some humans. Others are part of the co-production efforts of their users, such as in the way that a legally-available sex robot can often be modified to become a creepy and malicious child sex robot. Claude Shannon (1949) was one of the pioneers of the digital era, inventing some of its basic concepts such as digital circuit design and information theory. He made the following projection: "I visualize a time when we will be to robots what dogs are to humans, and I'm rooting for the machines." The possibility that humans will have a low social place in relation to robots relates to another of the major themes of this book (outlined in the next chapter), that of efforts to construe robots as "outclassing" humans in substantial ways (Moravec, 1988). The combination of robots often acting in rogue and unpredictable ways and humans themselves as feeling significantly outclassed signals critical social and economic problems for societies (Oravec, 2019).

© The Author(s), under exclusive license to Springer Nature Switzerland AG 2022
J. A. Oravec, *Good Robot, Bad Robot*, Social and Cultural Studies of Robots and AI, https://doi.org/10.1007/978-3-031-14013-6_1

Questions of how and whether robot developers, manufacturers, distributors, and implementers will exacerbate these concerns or even attempt to gain from them opportunistically will be explored in this book.

Even when robots are functioning well for a while, there are growing uncertainties: will the robot be hacked by someone, and not perform its intended duties? Will I understand how the robot functions, or will "dark patterns" lead me to interacting with the robot in a way that disadvantages me? What should I do when my own efforts are being compared to those of a robot in a format that disadvantages me? Who is responsible if the robot or autonomous vehicle is involved in an accident? Answers to these questions are emerging through political and legal actions and movements as well as social and personal interactions. Answers are also emerging as economic conditions foster more intense and higher numbers of interactions with robots and other autonomous entities; many of these interactions will be compulsory, as robots patrol our communities and become our co-workers. Social media and word-of-mouth are also playing roles in spreading and proliferating some of the positive as well as unsettling reflections people have about robots and AI. In the ensuing discourse, our basic concepts of what it is to function as a human being in a particular environment are changing.

WHO MIGHT BENEFIT FROM THIS BOOK?

This book will have some practical uses for marketing, management, educational, and technological development initiatives involving robotics and AI. With the emergence of robotics and AI, a tremendously powerful and malleable set of technologies is providing many kinds of opportunities for dark and creepy manifestations as well as for useful and functional initiatives. People are becoming frightened, anxious, and humiliated by robots and other AI-enhanced entities as well as using them capably for entertainment and various workplace functions. Some individuals are becoming romantically involved with them. Increasingly, we are being compelled if not forced by corporate and governmental organizations to deal with robots and AI entities as part of our everyday life. Many of the reflections in the book may seem to be attempts to diminish the hubris of developers and researchers that they can direct what users will eventually do with their products; it relates cases of individuals manipulating and abusing robots in ways that developers may not have had in mind. The book tells a complex tale of how user co-production combines with developers' own efforts to play large roles in high-tech arenas involving intelligent entities. In other places, the book can seemingly have a kind of diatribe against robots and AI with its emphasis on the negative, in contrast to many approaches in which the positive aspects are projected and emphasized (such as Donhauser, 2019). Some of the negative aspects of these entities may indeed be used to some construed opportunistic advantage in implementation and marketing efforts. It is hard to write material about the negative aspects of large-scale intelligent systems developments without sounding a bit conspiratorial and even anti-technological. As an example of such a conspiratorial approach in another field, researchers, reporters, and legal analysts have worked to unearth apparent collusions and schemes among experts who were entrusted with control the proliferation of opioids and related drugs in the US and many other countries (Ausness, 2018). As in the case of opioids, sometimes conspiracies are indeed identified and mapped in useful ways, leading to social and legal remedies.

I am hoping that someone who has concern and anxiety about robotics, autonomous vehicles, or AI will gain something from this book, if only to learn that others are facing comparable problems and anxieties. Whatever our levels of technical knowledge, we are increasingly called upon to deal with stressful issues concerning robots and other autonomous entities, with new kinds of economic challenges and social decisions often emerging. Our jobs could incorporate the use of an industrial or service robot, or our commutes can involve riding in an autonomous van. Our families may be considering the purchase of a robotic lawn mower or companion robot. We may use floor-cleaning or lawn-mowing robots and perhaps have become quite fond of them; studies show that more than 90% of Roomba users have a nickname for their device (Zimmermann et al., 2021). However, we may be worried about how the lawn mower is affecting wildlife around our home (Rasmussen et al., 2021). Specific social instances involving robots can leave us with lingering questions: for example, my late aunt died a couple of years ago at the age of one hundred; I had brought a selection of robotic toys during her last days, which I thought would be fun for her and her care assistants to play with. She hid the toys away. Was my aunt traumatized by the toys? I believe that my overall relationship with my aunt was rather good, and other gift items were not hidden away in a comparable manner. Unfortunately. I found out about her hiding the toys in a distant closet too late to get her direct feedback. Many researchers are conducting research studies on issues of consumer engagement and satisfaction with robots, however (such as Delgosha & Hajiheydari, 2021), so my anecdote about my wonderful aunt provides only one piece of a complex and growing puzzle.

This book might also be of value to those who are struggling to manage workplace and community situations involving robots and AI that are disconcerting and perhaps a bit hard to convey. For example, I gave a talk about service robots at a technical conference a while back. I was confronted in the middle of my exposition by an administrator who stated that the delivery and food dispensing robots I described would not work at his college. "The students will kick and bully them. They will not last long," he stated bluntly. Many of the other audience participants agreed with him. My counter-arguments about young people enjoying and appreciating robotics did not change anyone's position during this discussion. Although these two recent anecdotes in this preface cannot be taken as anything other than one person's narrative, the many other cases of misapplication and abuse discussed in these pages may provide more reason for concern. This book explores the prospects for an "anti-robot rebellion" of sorts or other forms of "rage against the machine." In the near-term future, such phenomena will take their places more fully alongside forms of human-to-human bullying, manipulation, and abuse.

THE DARK SIDE, CONSPIRACY, AND SCIENCE FICTION

Calling someone a "robot" is generally considered an insult in contemporary US culture (Rich, 2021), with connotations of inhumanity and emotional distance. For example, clerks serving coffee at restaurant chains have often been identified as "coffee-making robots" (Sainato, 2021), receiving increased levels of abuse along with this designation. Some nurse trainees are reported as stating "my greatest fear is becoming a robot" (Clinton et al., 2018). Psychologist and Holocaust survivor Erich Fromm wrote in *The Sane Society* about the perils of individuals becoming robots, comparing that status to those of human slaves (Fromm, 1955). More recent books have also been written with the theme of distinguishing humans from high-tech entities, such as Jaron Lanier's *You Are Not a Gadget: A Manifesto* (2010). Fearsome robots with unsettling and even disgusting traits are common on our television shows and movie screens. Why, then, would we want to work with them and have them in our homes? Why would some individuals state that we as humans are being "outclassed" by robots in many ways, as discussed in the first chapter (Moravec, 1988, p. 100)? The contradictions and inconsistencies involved are critical to our understanding of robots and other autonomous entities. The efforts of marketers to ameliorate upsetting issues concerning robots may not be sufficient to eradicate the incongruities.

Robotic initiatives have filled the thoughts and dreams of many individuals for centuries, from Leonardo da Vinci's early formulations (Moran, 2006; Rosheim, 2006) to the artistic "Wall-bot" installations of recent years (Ferguson et al., 2017). The dark side of robotics and AI has many dimensions, as outlined in the chapters to come; they include the intentional uses of the entities in dysfunctional ways to some inherent "creepiness" that may not dissipate as the years go on. The terms "dark side" and "evil" may seem a bit harsh, but as we examine the academic investigations, news accounts, and research scenarios of how robotics and AI are affecting modern societies in terms of military as well as everyday uses, they often seem appropriate. Expressions of empathy toward robots are also commonplace, as in the years-long discourse on an artist's video distributed in social media portraying a tired-looking robot who is consigned to clean up a nasty mess (Fowler, 2021). How we view robots and AI and integrate them into our societies reveals a great deal about ourselves as human beings as well as the character of our various social structures. Our discourse about bringing robots and autonomous vehicles into our lives is intersecting with personal and societal concerns about the integration of our fellow human beings into workplaces and communities, with themes of differences, migrations, capabilities, and disabilities. Some of our organizational and societal discourse about gender, racial, ethnic, and disability issues may be coded in robot- and AI-talk. For example, the tired-looking floor-cleaning robot video was originally intended to convey artistic messages about migration, surveillance, and border control, though the discourse expanded into exchanges about the place of robots in the social spectrum as well.

This book is not an attempt to project conspiracies and diabolical schemes on the part of individuals who are creating and implementing robots and AI-enhanced entities. Rather, the book attempts to identify trends and forces that may be shaping the use of robots and AI in dysfunctional ways. For example, what people can do and what robots can do may need to be compared in some way, but setting up unfair human-robot "contests" for marketing or demonstration purposes may damage human wellbeing and not give us the answers we need about how to implement various processes. Also, some of the humor that travels with robot and autonomous vehicle issues may be entertaining to some people but could direct attention away from critical issues of child sex images, misogyny themes, racial bias, and accidental deaths; this may especially be the case with sex and killer robots (discussed in separate chapters in the book). I hope this book serves to inspire those who have worked with robots and AI and who have done many exciting projects to examine the social and ethical ramifications of their efforts and listen to those who are affected by their initiatives. Openly proclaiming a fear of robots or a concern about autonomous vehicles is currently not conducive to thriving in modern workplaces (McClure, 2018), so there may be many other emerging robotics and AI-related issues that have not been widely discussed and have yet to be fully examined (and are only hinted at in science fiction).

Writing about robots and AI futures indeed begins to trespass into the realm of science fiction. For example, the work of Isaac Asimov (1940) set the direction for thousands of philosophers, public policy analysts, and interested others in exploring robot-related ethics, and Neal Stephenson's (1992) *Snow Crash* inspired some of the concepts behind today's "metaverse" initiatives. Nobel laureate Doris Lessing declared that "Science fiction is the dialect of our times" (Garnett & Ellis 1990). Donna Haraway (1987) states that "contemporary science fiction is full of cyborgs—creatures simultaneously animal and machine, who populate worlds ambiguously natural and crafted" (p. 2), discussing a large range of social constructions that include robots and AI-related entities. Many reflections and images from science fiction have played important roles in framing questions such as whether robots and other AI entities can co-exist with humans in peaceful, productive, and non-violent ways (Telotte, 2016). The long-standing legacies of science fiction can inspire other provocative questions, though: if the technology behind robots and AI is so effective, then why are many individuals still reluctant to have robots and autonomous entities assisting them safely and securely in their everyday lives? Such frightening (though generally comedic) films as *Big Bug* (2022) in which a 2050-era "group of bickering suburbanites find themselves stuck together when an android uprising causes their well-intentioned household robots to lock them in for their own safety" provide us with warnings as to what our own domestic life may look like. These imagined futures may perhaps forestall some of the more dire consequences to come. In the book *Silently and Very Fast* by Catherynne M. Valente, the following remarks from a robot signal strong intentions:

> I am not a Good Robot. To tell a story about a robot who wants to be human is a distraction. There is no difference. Alive is alive.
> There is only one verb that matters: *to be*.

As Valente's brilliant science fiction writing outlines, a robot may indeed not want to be "good."

The fact that the AI-enabled future that many of us envisioned years ago is still a way off signals that there are some significant social and ethical issues that need to be resolved about the technologies that are projected to be so valuable. Most of us still live in primitive, unsustainable homes that are difficult to cool, heat, clean, and otherwise manage as well as participate in automobile and public transportation systems that kill many drivers, passengers, and onlookers every year. We often still mow our own lawns and vacuum our own floors, despite the availability of robotic lawn mowers and household vacuums. The stories that are we telling ourselves and are being told about the potential dangers of robots and AI can influence how we select and use these technologies. Bartneck (2004) states that "some concepts about human robot interaction that have been promoted by movies and literature... utilize people's fear of the unknown to build engaging stories" (p. 35). The dangers of not recognizing the continuing power of these fears are discussed in this book, especially in the context of military, police, and security robots.

AI, CPSR, AND THE SANDBOX

Along with robotics, this book also discusses the dark sides of artificial intelligence (AI) as well as some of its more exciting and generative aspects. Decades ago, AI was a renegade science, and it also had strong creative and even ludic aspects. It was nearly always at the margins; when a computing approach became domesticated and widely utilized, it was often no longer considered "AI" (Bollier & Pavlovich, 2008; Oravec, 1996). For example, when I started to teach and do research in AI in the early 1980s, the notion of "object-oriented programming" was still largely attached to AI rather than being part of standard programming practices, as it later became.

Accepting and even glorifying failures is still an integral part of the AI mindset (Oravec, 2019). Whole fields of AI were branded as failures but later became prominent (such as machine learning). Decades ago, I organized a special issue of the *Journal of Systems and Software* on the varieties of failures of expert systems and what we can learn from them (Oravec & Plant, 1992; Oravec & Travis, 1992). However, when AI approaches were injected into mission-critical applications, such as systems that are relied upon to detonate enemy missiles before they reach their targets, new and very serious kinds of concerns about failure and playfulness emerged. As AI matured, research money became easier to obtain; playful attitudes toward computer science are often harder to come by when commercial and military grants are on the line and professional statuses demand something labeled by others as "results." CPSR (Computer Professionals for Social Responsibility) members often used the playful "sandbox" notion in discourse; CPSR gave its members like myself a sandbox in which to experiment with their nascent ideas about computing and social issues. The sandbox notion is indeed a common one in computing as a whole (Hopper, 2018), so its use was not an insult but an apt image related to creative thinking. I hope that this book and the discussions that follow will have some of the generative feel of the sandbox, although I have done my best to tighten arguments when needed and provide relevant sources that will empower people to do their own subsequent investigations. Although the CPSR no longer exists as an organization, its impacts on its members have been long lasting, as related by Douglas Schuler (1994) and many others. I will discuss the CPSR as a pioneering organization further in the chapter on military and police robots.

References

Ausness, R. C. (2018). The current state of opioid litigation. *South Carolina Law Review, 70*, 565–607.

Bartneck, C. (2004). From fiction to science – A cultural reflection on social robots. In *Proceedings of the CHI2004 Workshop on Shaping Human-Robot Interaction* (pp. 35–38). Vienna, Austria. https://doi.org/10.6084/m9.figshare.5154820

Big Bug. (2022). Eskwad and Gaumont Film Company. Director Jean-Pierre Jeunet (Distributed by Netflix.)

Bollier, D., & Pavlovich, R. (2008). *Viral spiral: How the commoners built a digital republic of their own* (p. 156). New Press.

Clinton, M., Madi, M., Doumit, M., Ezzeddine, S., & Rizk, U. (2018). "My greatest fear is becoming a robot": The paradox of transitioning to nursing practice in Lebanon. *SAGE Open, 8* (2). https://doi.org/10.1177%2F2158244018782565

Delgosha, M. S., & Hajiheydari, N. (2021). How human users engage with consumer robots? A dual model of psychological ownership and trust to explain post-adoption behaviours. *Computers in Human Behavior, 117*. https://doi.org/10.1016/j.chb.2020.106660

Donhauser, J. (2019). Environmental robot virtues and ecological justice. *Journal of Human Rights and the Environment, 10*(2), 176–192.

Ferguson, S., Rowe, A., Bown, O., Birtles, L., & Bennewith, C. (2017, June). Networked pixels: Strategies for building visual and auditory images with distributed independent devices. In *Proceedings of the 2017 ACM SIGCHI Conference on Creativity and Cognition* (pp. 299–308).

Fowler, K. (2021, November 12). Why the Internet is feeling sorry for a robot: "Screaming and panic." *Newsweek*. https://www.newsweek.com/internet-feeling-sorry-robot-screaming-panic-installation-tiktok-1648626

Fromm, E. (1955). *The sane society*. Fawcett.

Garnett, R., & Ellis, R. J. (1990). *Science fiction roots and branches: Contemporary critical approaches*. Springer.

Haraway, D. (1987). A manifesto for cyborgs: Science, technology, and socialist feminism in the 1980s. *Australian Feminist Studies, 2*(4), 1–42.

Hopper, S. (2018). The heuristic sandbox: Developing teacher know-how through play in simSchool. *Journal of Interactive Learning Research, 29*(1), 77–112.

Ivanov, S. H., & Webster, C. (2017, June). The robot as a consumer: A research agenda. In *Marketing: Experience and Perspectives Conference*, 29–30 June 2017, University of Economics-Varna, Bulgaria (pp. 71–79).

Lanier, J. (2010). *You are not a gadget: A manifesto*. Alfred A. Knopf.

McClure, P. K. (2018). "You're fired," says the robot: The rise of automation in the workplace, technophobes, and fears of unemployment. *Social Science Computer Review, 36*(2), 139–156.

Moran, M. E. (2006). The da Vinci robot. *Journal of Endourology, 20*(12), 986–990.

Moravec, H. (1988). *Mind children: The future of robot and human intelligence. Cambridge*. Harvard University Press.

Nader, R. (1965). *Unsafe at any speed. The designed-in dangers of the American automobile*. Grossman Publishers.

Oravec, J. A. (1996). *Virtual individuals, virtual groups: Human dimensions of groupware and computer networking*. Cambridge University Press.

Oravec, J. A. (2019). Artificial intelligence, automation, and social welfare: Some ethical and historical perspectives on technological overstatement and hyperbole. *Ethics and Social Welfare, 13*(1), 18–32.

Oravec, J. A., & Plant, R. (1992). Guest editors' introduction. *Journal of Systems and Software, 19*(2), 111.

Oravec, J. A., & Travis, L. (1992). If we could do it over, we'd… Learning from less-than-successful expert system projects. *Journal of Systems and Software, 19*(2), 113–122.

Rasmussen, S. L., Schrøder, A. E., Mathiesen, R., Nielsen, J. L., Pertoldi, C., & Macdonald, D. W. (2021). Wildlife conservation at a garden level: The effect of robotic lawn mowers on European hedgehogs (Erinaceus europaeus). *Animals, 11*(5), 1–13. https://doi.org/10.3390/ani11051191

Rich, K. (2021). Equipment as living: Robotic rhetorical homology in humans. *Popular Culture Studies Journal, 9*(1), 117–137.

Rosheim, M. (2006). *Leonardós Lost Robots*. Springer Science & Business Media.

Sainato, M. (2021). Coffee-making robots': Starbucks staff face intense work and customer abuse. *The Guardian*. https://www.theguardian.com/business/2021/may/26/starbuck-employees-intense-work-customer-abuseunderstaffing

Schuler, D. (1994). Community networks: Building a new participatory medium. *Communications of the ACM, 37*(1), 38–51.

Shannon, C. (1949). Communication theory of secrecy systems. *Bell System Technical Journal, 28*(4), 656–715.

Stephenson, N. (1992). *Snow crash*. Bantham.

Telotte, J. P. (2016). *Robot ecology and the science fiction film*. Routledge.

Valente, C. M. (2012). *Silently and very fast*. Cape Town Books.

Zimmermann, J. L., de Bellis, E., Hofstetter, R., & Puntoni, S. (2021, November). Cleaning with Dustin Bieber: Nicknaming autonomous products and the effect of coopetition. In *TMS Proceedings 2021*. https://tmb.apaopen.org/pub/nrze79i7/release/1

CHAPTER 2

Dramaturgical and Ethical Approaches to the Dark Side: An Introduction

Robots and other autonomous entities have sparked curiosity, fear, and anxiety for decades. Are robots best construed as compliant and friendly companions and helpers? Or, are they better characterized as conniving, deadly, and menacing? Making generalizations and doing analyses in this arena can be difficult for researchers and developers: individuals' experiences with robots and other AI applications can vary dramatically. One person may have purchased a robot toy or an automated lawn-mowing device; for this individual, dealing with robots may be akin to having a new puppy, with particular annoyances yet occasional delights. In contrast, other people may have lost their jobs or had their employment radically changed in character in part because of the workplace implementation of robotics or autonomous vehicles. Still others may be concerned about the police applications of AI because of various robotic encounters during a protest. Perhaps others have encountered a possibly-fanciful announcement about upcoming robot ventures, such as the following:

> [Elon] Musk apparently is building a humanoid robot. It will be five feet, eight inches tall, weigh 125 pounds and have human-like hands. What will she/he/it do? Basically, the Tesla Bot will be our cyber slave, handling tedious and repetitive tasks. For example, you can tell it to go to Chipotle and get a burrito—and it will happen. (Taulli, 2021, para. 2)

This book addresses such developments and their related social changes, with an emphasis on the negativity that often permeates many robotic and autonomous entity interactions with humans. It explores both the immediate changes that robots and other AI-enhanced entities are making in our social contexts and physical environments as well as the unsettling conceptual and cultural inheritances that these entities often carry (including the vast assortments of robot-themed literature and films as well as decades of "AI hype"). It uses dramaturgical and ethical perspectives in attempts to gain some insights about how these emerging technologies are affecting deeply-seated social and psychological processes. Unlike some other high technologies that are rather novel in their recent origins, robots are associated with substantial narrative and image-related legacies that often place them in a negative light, adding more complexity to the mix.

The chapters to come describe a large assortment of useful robotics and AI applications for which very positive outcomes are generally expected and often promoted by marketers. However, the major emphasis of these chapters is on how these technologies are being abused and their dark sides manifested and often intentionally utilized in opportunistic ways, in order to explore fully some dimensions of these entities that are often hidden, repressed, lampooned, or ignored. The robot as an "other" in the workplace and community—an object of attention and discussion—has been a frequently-utilized theme of science fiction and creative work as well as a topic for research analysis for many decades. Another theme discussed in upcoming chapters is that many people are "acting out" their anxieties and grievances by physically attacking robots and autonomous vehicles or manipulating the software of these entities. Human-AI contests and displays of robotic feats are often used to intimidate workers and reinforce that individuals are not in control of their own employment destinies, which presents unsettling prospects for the future. Some of these contests are moving to community and domestic spheres as various developers and businesspeople attempt to show that robots outperform humans in battlefields and bedrooms. Robotics and other AI applications are also becoming powerful tools for organizations to influence how individuals see themselves and for police and military agencies to control and subdue civilians. The uncertainties involved in how these AI-enhanced entities operate have often become a feature rather than a flaw in their design as they spawn anxieties and fears for the public.

Robots and autonomous vehicles are already playing substantial roles in individuals' and organizations' performances of social and environmental mastery, for example as dangerous and isolated regions (including outer space) can be monitored and utilized remotely without direct human interaction (Wong et al., 2018). Robots and other AI-enhanced entities can be connected to the other applications people use in the metaverse, controlling virtual as well as real-life environments. People who question the expansion of the use of robotics in society can face strong opposition from robotics and AI advocates who preach their inevitability and try to defend their economic benefits, and who often assert that these entities "outclass" humans in some dimensions (Oravec, 2019). Projections of a future-world-with-ubiquitous-robots are growing, though the kinds of social and psychological preparations that should be made to support its arrival may not be adequate. As discussed in sections to come, the robot and AI-related "shocks" with which we may soon be faced may be comparable to any other kind of "future shock" that technological and social changes may engender.

Dramaturgical Perspectives on Robots and AI

A dramaturgical perspective (such as the one used in this book) is especially appropriate with entities that have roots in the theater. The name and some of the basic concepts involving robots are linked with a drama, the 1920s play *RUR (Rossum's Universal Robots)* by Karel Čapek (1923). *RUR* served to proliferate the term "robot" in its modern form and framed the robots involved as being artificial personages. Szollosy reflects on some of *RUR's* historical underpinnings:

> Čapek's play was staged in this period of increasing anxiety as to what was happening to us as a species, as a culture, in a period (i.e. the very late nineteenth, early twentieth century) and place (i.e. Western Europe) where there is much interest elsewhere in dissociative disorders (e.g. Freud), mechanisation (e.g. Frederick Winslow Taylor, F. T. Marinetti) and alienation (e.g. Marx et al.). Čapek's play reflects these concerns regarding *dehumanisation* in the light of mass industrialization, the increasing mechanisation of killing seen during The Great War and a loss of certain pastoral conceptions of human experience in an increasingly urbanised world of technological innovations. (Szollosy, 2017, p. 433)

The play had "immediate and significant literary influence" (Dinan, 2017, p. 108) that helped to stimulate discourse about the societal prospects for automata. Guizzo and Ackerman (2016) characterize the theme of "robots killing off the human race" as "an enduring plot line in science fiction: robots spiraling out of control and turning into unstoppable killing machines" (p. 38). Popular movies such as Fritz Lang's 1927 *Metropolis* (Elsaesser, 2012) and James Cameron's *The Terminator* (1984) have continued on the legacies of these often-negative yet dazzling and futuristic themes and images. Čapek's reflections inspired the consideration of the relative places of machines and people in the social order: "We have made machines, not people, the measure of the human order, but this is not the machines' fault, it is ours" (quoted in Klíma, 2002). Some early drone-related concepts were implemented in Nikola Tesla's work in 1898; he "controlled a boat on a pond in Madison Square Garden by means of a radio signal, the first such application of radio waves in history and arguably also the genesis of modern robotics" (Casey-Maslen et al., 2018).

Various robotic implementations subsequently emerged from the era of *RUR* and *Metropolis*. Bhaumik (2018) relates that the first humanoid robot followed quickly in this early period. "Eric" was reportedly made by W. H. Richards in 1928 and put on public display in London. On a comparable theme, the "Roll-Oh—the Domestic Robot" film was shown at the 1939 World's Fair, portraying a fictional robot:

> "Roll-Oh" can grasp objects, has a retractable knife in its hand, as well as a plant watering system, a can opener, and a gas-flame lighter. Its foot is also a vacuum-cleaner. Tongue-in-cheek film showing a domestic robot freeing housewives of their chores and intimating that their work is hardly necessary. (Phelan, 2017, from the Prelinger Archives)

The Westinghouse Electric Company's "Elektro" robot of the 1940s "was controlled by electrical relays and could play recorded speech, respond to voice commands, and move on wheels attached to its toes" (Bhaumik, p. 4). Elektro often appeared with "Sparko," a robot dog that could "bark, sit, and beg to humans" (Sharkey, 2008). The line between what was simulated (enacted by humans behind the scenes and backstage) and what was automated was often unclear in these presentations, as is still the case in some robotic demonstrations. Films that included rogue robots began to proliferate, often linking robots to alien worlds. For example, the 1953 cult classic *Robot Monster*, a black-and-white 3D film (produced and directed by Phil Tucker and distributed by Astor Pictures), recounts the story of the robot Ro-Man's mission to Earth to destroy humanity.

As in the early years of robotics, today's robots "give rise to hopes and fears about the future and our place in it" (Küster et al., 2021), with the demonstration of a robot's capabilities often stimulating societal discourse. Consider the dramatic November 2021 description of the unveiling and release of a new robot, called "Ameca":

> Meet Ameca: … the 'world's most advanced' humanoid robot has been unveiled in a UK lab Machine, named Ameca, has eerily realistic facial expressions and movements. It's been designed by British company Engineered Arts and revealed on YouTube (Tonkin, 2021, para. 1).

A short video portraying Ameca that was released by Engineered Arts was discussed in a number of newspaper stories and social media interactions: "At the start of the video, Ameca appears to 'wake up,' as its face conveys a mix of confusion and frustration when it opens its eyes. But when Ameca starts looking at its hands and arms, the robot opens its mouth and raises its brows in what it looks like is amazement. The end of the video shows Ameca smiling and holding a welcoming hand out towards the viewer—if that's how you want to interpret that gesture" (Roth, 2021). The fact that a short video release about a robot not yet on the market can attract considerable attention in investment circles as well as social media and newspapers demonstrates that we are still in a malleable stage of technological development, one in which people's reactions and perspectives can potentially still have substantial influence.

Most of this book pertains to workplaces and community settings, not drama and the theater directly; however, the powerful concepts and insights that are linked with dramaturgical approaches (such as "frontstage" and "backstage" perspectives) can help to clarify some nuances of human-robotic interaction (Eckersall, 2015). For instance, the frontstage appearance of the robot Ameca is designed to correspond to humanlike features: "The company says that it used 3D scans of actual humans to give the bot accurate bone structure, skin texture, and lifelike facial expressions" (Roth, 2021), although many other robots have less of a humanoid appearance. Some recent theatrical performances have indeed integrated robotics in central ways: "we look at the use of robots in recent theatre and argue that, in some cases, robots are ideal vehicles for performance based on new media dramaturgy as they can translate between the informatic and the organic, facilitating meaningful transactions" (Eckersall et al., 2017, p. 107). Nijholt (2019) and Vilk and Fitter (2020) describe efforts to include robots in standup comedy routines, underscoring the difficulties of timing and adaptivity in these complex performative environments.

Dramaturgical perspectives have been extended in a variety of dimensions into high-tech arenas. For instance, Erving Goffman's (1978) work on performances and the "presentation of self" in everyday life has played pivotal roles in how the social impacts of computing have been characterized for decades (Bullingham & Vasconcelos, 2013). Goffman's research attempted to capture.

> ... how individuals 'perform' in order to project a desirable image, using the theatre to illustrate individuals' contrasting front stage and back stage behaviour. During interaction, those participating are viewed as actors... When in front stage, an actor is conscious of being observed by an audience and will perform to those watching by observing certain rules and social conventions, as failing to do so means losing face and failing to project the image/persona they wish to create. (p. 101)

With humans, frontstage "leakages" of various physiological cues (especially relating to the face and hands) can convey information about "backstage" processes; individuals who endeavor to lie have often been foiled in their efforts by sweaty palms, wringing hands, or shifting eyes. Such leakage can apparently signal increased levels of anxiety on the part of an individual relating to a particular topic, possibly relating to lying (Burgoon, 2019; Denault & Dunbar, 2019; Oravec, 2022). Robots do not have the same kinds of frontstage leakage, however, and thus, the clues that keep us as humans in synch with each other and often trusting each other at some level are missing. The fostering of "overtrust" that characterizes some of the interactions between humans and robots (Ullrich et al., 2021) may be rooted in this lack of robotic leakage. If we could see robots "leak" in some authentic way (as we do when a human is involved), we may be able to develop more effective relationships with them.

Dramaturgy can provide us with other insightful perspectives dealing with emotional manifestations. "Dark patterns" are being developed by designers and implementers that can lead users astray in their interactions with technologies; these backstage manipulations present themselves as involving a certain, specific set of straightforward interactions but are designed to have hidden manifestations that put users at a disadvantage (Luguri & Strahilevitz, 2021). Examples of these are provided in the chapter on sex robots, including the surreptitious elicitation and collection of personal information. In many dramaturgical approaches, the audience can be manipulated or deceived in some way, so dark patterns strategies are in keeping with a long theatrical as well as social science tradition (Tseëlon, 1992). Strong emotions on the part of users can make these dark patterns less noticeable (more opaque), as the users are distracted. Expression of emotions concerning robots and AI can swing wildly, augmenting the emotional content and dramaturgical value of various real-life situations. There are certainly some primitive "I hate robots" sentiments expressed at various times and places (as outlined in the robot abuse analyses to come); the US performer Ice Cube's anti-robot statements (Nas, 2020) are one example (if they are indeed not entirely parodies). In a positive direction, the coupling of a robot dog with the South Korean BTS band brought international attention for Hyundai's robotic initiatives (Gartenberg, 2021). Considerable affection for robots in the Japanese culture has also been well documented (Knox & Watanabe, 2018; Sone, 2016).

There are many complex and nuanced expressions about robots and AI that are emerging in books, articles, films, and social media postings. Philosopher Amit Ray (2018) asserts that "The coming era of Artificial Intelligence will not be the era of war, but be the era of deep compassion, non-violence, and love." The possibility that many individuals are aspiring toward some kind of spiritual or mystical connection with robots and AI through sexual communion and marriage is also a growing factor in the legal and ethical consideration of robotics (Yanke, 2021). In some cultures, the deaths of certain robots have been treated in a manner comparable to the death of a living entity (White & Katsuno, 2021) with "robot pet memorial services" emerging (p. 222). The term "lovotics" was coined by David Levy in an effort to encapsulate the various aspects of human love for robotic and AI-enhanced entities (Cheok et al., 2017; Levy, 2009). "Love" here is construed as a "contingent process of attraction, affection and attachment from humans towards robots and the belief of vice versa from robots to humans" (Samani, 2012). Dubé and Anctil (2020) along with others have used the term "erobotics" to refer more specifically to the use of robots and AI as erotic agents (as discussed in Chapter 3). Tendencies to hate robots and other AI-enhanced entities have not yet been as well framed and specifically characterized as those of robot love, but may eventually become larger and more tightly distinguished factors as anti-robot sentiments and activities intensify.

Since this book focuses on the dark and creepy sides of robotics (as well as other robotic and AI influences), it will examine how robots and other autonomous entities can behave in predictably dark and creepy modes as well as in unpredictable, "rogue" ways. Examples of the latter are how robots and autonomous vehicles can be manipulated and controlled by saboteurs or hackers and how the entities' programming can elicit inappropriate results. Sometimes a frightening or even terrifying demeanor can indeed be appropriate for a robot if it is designed to scare away trespassers in a toxic or otherwise dangerous environment. However, it is somewhat problematic for robots to inspire fear and anxiety intentionally in various other confined settings, although such initiatives are emerging especially in military, police, and everyday security contexts. An emerging set of "artificially constructed social signals and accommodating human expectations" (Hoffman, 2011, p. 2) has accompanied the use of robots and other AI-enhanced entities in various contexts, which provide a great deal of information about how human–robot interactions will transpire. As human experience with robots increases, these expectations may produce different iterations and related nuances.

Acquiring a Taste for Creepiness?

Audiences and markets for products and services that have a "dark side" theme can be groomed in various ways; horror stories in film and literature have acquired some willing audiences, for example, with "horror" defined as fear of some uncertain threat to existential nature and disgust over its potential aftermath (Weaver & Tamborini, 2013). The sizable market for the horror movie genre demonstrates that creepiness has an audience (McKenna & Proctor, 2021); many of these films involve robotics and AI. Through advertising and social media, building such a market for robots and AI (however creepy and fearful they appear for some individuals) may be feasible. There are some recent technical demonstrations that explore the scary dark sides of AI; for example, Shelley, billed as "the world's absolute first community oriented AI repulsiveness story writer," co-wrote more than 200 horror stories collaboratively with humans, reportedly by learning from their nightmare ideas and expressions (Kumar et al., 2019).

Whether a "taste for creepiness" in the robotic, AI, and automated vehicle realms can become widespread in particular cultural settings has yet to be determined. Decisions about the use of sentient robots in Disney theme parks in the US demonstrate how difficult it can be to develop robots that are palatable to a variety of individuals, with different assumptions about robots (and tolerance for creepiness):

> The development of Groot—code-named Project Kiwi—is the latest example. He is a prototype for a small-scale, free-roaming robotic actor that can take on the role of any similarly sized Disney character. In other words, Disney does not want a one-off. It wants a technology platform for a new class of animatronics. Cameras and sensors will give these robots the ability to make on-the-fly choices about what to do and say. Custom software allows animators and engineers to design behaviors (happy, sad, sneaky) and convey emotion. (Barnes, 2021)

The choices that Disney Corporation makes in presenting its robots in "frontstage" configurations, along with the comparable choices made by other organizations and individuals, are shaping how robots and other AI-enhanced entities are being construed by broad consumer audiences. The book addresses the social responsibility of the individuals and organizations that are profiting from creating and developing robotics and AI technologies, but also of the implementers and users whose modifications and adaptations of the robots and AI-enhanced entities refashion and expand their functions. It hopes to inspire people to consider the social issues concerning autonomous entities in a visceral way as they examine for themselves the creepiness, anxieties, and addictions associated with them. Should the developers who are considering what robotic imagery will make a robot more frightening or an autonomous vehicle more imposing halt their efforts? How should researchers who are exploring what sex robot colors and sounds will make them more addictive to users proceed in their research? By conducting such research, we indeed learn more about human beings and their tolerances for technological impositions and disruptions, at the cost of considerable human discomfort. As a society, we are learning how much pain can be caused by technologies before humans organize and rebel. We are also finding out whether the children that are playing with and being taught by robots rather than humans lack something important in their lives. More critical and enlightened approaches to these robot-related research initiatives may indeed result in valuable insights about how we relate to each other as humans and how we can make life more tolerable for everyone.

Disturbing images and disquieting robotic narratives have had a continuing legacy in many cultures, often intensifying the impacts of the robotic and autonomous vehicle abuses and disturbances that occur. Consider the robot Hal in *2001: A Space Odyssey* (Kubrick, 1968), whose autonomous activities put the entire mission into a tailspin; as the spaceship's computer, Hal decided that it had to kill all the humans aboard because "this mission is too important for me to allow you to jeopardize it." A plethora of science fiction movies or narratives with some levels of robotic destruction (destruction by and of robots) is a part of the cultural infrastructure of robotics and AI. Also, a part of many robot-related narratives is discussions of whether individuals' jobs are likely to be displaced or drastically changed in character (Autor, 2015), a kind of disruption that can often be as frightening as physical threats to those who sustain it. Some attacks by humans against robots and autonomous vehicles may be related to these upsetting images (as described in upcoming chapters) although other acts of mayhem may be expressions of various kinds of personal disquiet or psychological turmoil. Some of those who throw eggs at autonomous vans (as described by Un Nisa, 2021) could very well also throw eggs at more directly human-controlled, traditional vehicles. Dissemination of narratives about misuse of self-driving vehicles' autopilot functions is distributed on social media sites; Bindley and Elliott (2021) describe how "some [Tesla] users brag online about how they misuse their cars by tricking the driver-assistance system."

As outlined in the following pages and in a great deal of related research, the streams of narratives and images produced in popular discourse on such entities as sex robots and police and military robots are shaping perceptions of AI-related initiatives (Devlin, 2018; Krishnan, 2016). For example, titles of newspaper articles such as "Doctors 'need to prescribe sex robots to cure digisexual illnesses,' experts urge" (Edwards, 2021) are attracting public attention and expanding people's imaginations about emerging robotic and AI developments as well as the related activities of experts. Simply put, people are buying and using the sex robots described, expanding the market for the products. Linkages between sex robot usage and human slavery can also engender disturbing connections that could exacerbate unfortunate human foibles (Hampton, 2015; Richardson, 2016); connections of human beings with slavery apparently never end well in modern societies. The primary purposes of these sex robot narratives by their creators and disseminators are generally not to contribute to academic research efforts such as this book; rather, many of the narratives are designed to attract people's attention as "clickbait" or to influence their behavior in using robots. For many other robot narratives, the purpose is to sell robots and in the case of police and military robots, to inspire fear and motivate others to cooperate with the robots' orders. As robots become more widely integrated into many kinds of applications in homes, many new narrative themes are likely to emerge with more domestic and personal aspects, possibly related to science fiction stories or marketing materials.

The dark sides of other kinds of technology, and how they have been constructed and contained through the decades, can provide us with some insights as to how to proceed. Many new technologies have been influenced in their development by negative images or dystopian themes. For example, the nuclear power industry in the US was affected by the "China syndrome" themes of movies and print media in the 1970s and 1980s, in which the immanent destruction of some nuclear power station facilities was predicted (Shaw, 2013). A selection of film and TV references as well as popular and tabloid journalism pieces is included in the chapters to come in order to frame the characterizations of robotics and AI in everyday discourse (such as those provided in Sundar et al., 2016). Szollosy (2017) describes some of the robotic figures and monsters that have emerged as reflecting a perception that human nature has itself been dehumanized through its scientific advancements and technological changes over the past two hundred years: "Using the psychoanalytic notion of *projection*, these monsters are understood as representing human anxieties regarding the dehumanising tendencies of science and reason" (p. 433). Reportedly, along with the negative and dystopic angles have been very positive, "godlike" characterizations of robots, in which the supposed superiority of these technological entities is manifested.

Ruminating on the Prospects for Robots and Other AI Entities

Some of the prospects for robotics and automated vehicles in society were largely introduced decades before actual functioning items were made available for widespread usage. For instance, mockups and prototypes of robots in fiction as well as in industrial system demonstrations often described robotic characteristics before the functioning entities were available for usage. The "rumination time" that enables society to think about the implications of these technologies has both positive and negative dimensions. Various assumptions and expectations are wrought during this early period that can affect the social and ethical appropriateness of the applications to which the technologies are put. Some military and police applications of robots can be "normalized" before they are actually put into practices on a widespread basis, for example. Ruminating about the technologies can elicit inappropriate fears and misconceptions as well as engender overly optimistic visions of productive uses.

Ruminating about the prospects of robotics and AI has pulled humans in various directions. Research has shown that people can both feel very strongly for and against various entities, as is the case for certain celebrities and political candidates (Huddy et al., 2008). The "dark side" of robotics and other AI-enhanced entities comprises many different aspects, including human-to-robot violence, fear, anxiety, and deception. There can be a strong dark side connected with a particular technology as well as a brighter, more functional side, running in tracks that are generally parallel but which can often intersect. This book focuses on such dark side matters as the actual instances of robot to human violence, attacks and abuse of robots by humans, as well as the use of robots to foster a diffuse sense of fear in police and military contexts. Human interactions with robots are likely to vary widely in character for at least the near future. Some people will continue to vent their frustrations toward robots and autonomous vehicles in violent acts, or participate in forms of technological sabotage. However, many people will continue to adopt robotics and AI into their everyday routines, often with little awareness or concern about potential negative results. Corporations that want humans to hoard or become addicted to AI-related entities in order to increase sales, and agencies that exploit the fears that robots can engender, will have many opportunities to accomplish their goals as individuals become less critical, more habituated, and less aware of the roles of these technologies in their lives. Those who trust blindly the robots and other AI-enhanced entities with which they work and play may suffer serious consequences. Potential security breaches can make any robot untrustworthy, changing its character from a helpful agent to an attacking menace.

Robotics and AI are not the only technical arenas that have a kind of dark side. Studies of negative technological impacts have been focused on social media trends (Baccarella et al., 2018) and Internet gaming (Hyrynsalmi et al., 2017), as well as online academic interactions (Oravec, 2019). The cumulative impact of these negative technological alterations on our environments and everyday activities can put stress on our already overstressed systems. Tarafdar et al. (2013) declare that "we may be entering an era in which human frailties begin to slow down progress from digital technologies" (p. 61). Beck (2020) and others have identified the prospects for the "infantilization" of many of our mental processes, making us ill-equipped to face these challenges: "Technology could free us from ruminating on existential decisions and greatly simplify everyday conduct by providing a ready-made belief system as well as standard options of action" (p. 33). The kinds of warnings about our abilities as humans to face these challenges were echoed decades ago by futurists: anthropologist Margaret Mead (1965) described "many kinds of disjuncture" that were occurring in culture as technological, social, and educational changes proceeded. Futurist Alvin Toffler (1970) also wrote decades ago about a "future shock" that could disempower our attempts to deal with massive and consequential societal changes. Douglas Rushkoff's (2014) *Present Shock: When Everything Happens Now* reflected on the timing of these societal changes. A kind of "robot shock" can occur when our mental processes are overloaded in our interactions with robotic technologies, challenging our technological competences as well as our social and psychological assumptions and mooring. Individuals who are acclimated to a world of individually-functioning robots may be somewhat unsettled by an environment saturated with robots that are interconnected with sensors and processors in an "Internet of Robotic Things" or IoRT (Simoens et al., 2018). IoRT operates at a level of analytical density that humans cannot comprehend, although a brain–computer interface that utilizes electroencephalographic signals may provide some connections to individuals (Sudharsan & Deny, 2021). Mead's, Toffler's, and Rushkoff's perspectives as well as the "robot shock" insights can alert developers, implementers, and users to be critical and vigilant to the emerging directions of new technologies, rather than succumbing to their siren calls.

What is to be done in the wake of the technological and social changes? The final chapter of this book provides an exploration of how to be authentically human in a world in which we are increasingly being shaped and nudged by robots and AI entities and narratives, often in ways that are not in our long-term best interests, sometimes through "dark patterns" that are injected by developers into our interactions with technologies. What is striking about each of the topics of the preceding chapters is how inevitable each of the dark side concerns can appear: for example, there will undoubtedly be many more deaths and injuries linked to robots and autonomous vehicles. The chapters on the growing proliferation of sex and police robots are comparably imbued with a sense of inevitability, despite the best efforts of initiatives such as the Campaign Against Sex Robots (founded in 2015) and the Campaign to Stop Killer Robots (formed in 2012). The coronavirus pandemic served to reinforce the inevitability of the assumption that robotics would be required in order to fulfill the basic needs of society, largely because of the labor shortages in many sectors of economies (Brakman, Garretsen, & van Witteloostuijn, 2021). However, the last chapter describes some resistance against the "inevitability" of robotic and AI revolutions in society; the growing number of efforts of some researchers in high-tech arenas to be proactive against specific technological applications that run counter to human rights provides optimism.

Dissecting Robots: An Advisory About Potentially Upsetting Material

A cautionary notice is in order: as just stated, the book examines some of the disconcerting and dysfunctional implications of interactions with robots and other AI entities as individuals lash out against them or become disempowered by them, so some of the material may be unsettling. For some individuals, the discussion of "sex robots" will be unsettling; for others, the very idea of "killer robots" will elicit disturbing images of out-of-control sequences leading to an apocalypse. The horrifying cases of individuals crushed by robots while they worked in their everyday jobs or were killed when struck by an autonomous vehicle will stun many of us; we should all remember and revere the names Wanda Holbrook, Robert Williams, and Elaine Herzberg. (Their cases will be discussed in an upcoming chapter.) Some of the new science fiction treatments of robots and AI (such as the television series *Black Mirror*, created by Charlie Brooker) will be unsettling for some viewers; for me, Fritz Lang's 1927 *Metropolis* can still provoke nightmares with its portrayal of mob violence against the robot and the main human character. Lang's good robot/bad robot character structure in the 1927 film *Metropolis* makes strong points about the ills of mechanization and the potentials for dehumanization: "creature made to resemble living woman Maria looks like she, but 'inside' is her opposite. Maria-woman embodies Christian values of non-violence, patience and belief. But her double, the robot Maria cases revolution of workers, chaos and destruction. Lang presents the robot as an embodiment of evil, and as an instigator of the dark side of human character" (Horáková & Kelemen, 2006, p. 28). However, in contrast to the negative, evil-appearing robotic exemplars of this early era is the previously-mentioned Westinghouse Electric Corporation "Elektro," an upbeat and friendly mechanical robot "who performed a host of tricks to entertain visitors" (Terzian, 2009, p. 901; "Visitors to N.Y. Fair," 1939).

The kinds and perspectives of these negative narratives of robotics and AI are growing, filling tabloids, popular media, and word-of-mouth communications along with research studies. Some of these adverse accounts reflect the many all-too-real robot-inflicted deaths and AI-related damaging incidents. However, others are compilations of pieces of various myths and visions or fabrications of new fictions concerning robots. Robots and other AI entities are entering workplace environments that are already facing stresses from considerable economic and social disruptions, so the negativity in these narratives and images could be amplified or reshaped in unsettling ways. Some robots are indeed part of the surgical process of dissecting humans and other animals (Lira & Kowalski, 2020). Robots are themselves dissected for many developmental and research reasons: the very act of dissecting a robot can be upsetting (one reason for the warning for readers). Just seeing the insides of a machine is rarely alarming to people; we open the case of a flashlight to put in batteries or inspect what is going on in our washing machines by removing panels. However, strikingly-compelling robot dissections are commonplace in literature, film, and education, often providing considerable dramatic value. Kakoudaki (2014) states "Stories and films featuring robots, cyborgs, androids, or automata often stage scenes that depict opening the artificial body: someone ejects a face plate, pulls back artificial skin, removes a skull covering, reveals a chest panel, lifts clothing, or pushes a button, thereby rendering visible the insides of the fascinating human-like machine" (p. 1). Dissecting a robot can be disconcerting to watch, especially when the face is peeled away; the "Frubber" fake skin that provides the facial foundation for many robots has been developed to look like human skin and the parallels to human dissection are strong. The term "frubber" is "a contraction of flesh and rubber"; the material is designed to "wrinkle, crease, and amass, much more like skin" (Oh et al., 2006, p. 1433). For metaverse developments (described in the next chapter), new and more responsive forms of robotic skin and related sensors are being developed by Meta Corporation and its collaborators:

> The thin 'robot skin', called ReSkin, uses machine learning and magnetic sensing to 'feel' objects it comes in contact with in a way that is similar to how humans interact with objects. Up to 3mm thick, the material was created in collaboration with researchers at Carnegie Mellon University. "ReSkin offers an inexpensive, versatile, durable and replaceable solution for long-term use…" "It employs a self-supervised learning algorithm to help auto-calibrate the sensor, making it generalisable and able to share data between sensors and systems. "The technology was shown to be able to train robots to perform complex actions such as unlocking a door with a key or grasping delicate objects such as berries and grapes with just the right amount of force, as the company demonstrated in GIFs in its blog post. (Gain, 2021, para. 4-5)

The intricate details of robotic skin may seem to be arcane, yet they are apparently playing important roles in how sex robots are being constructed and marketed. Consider the following description of the detailing of EX Doll sex robots (further explored in a later chapter):

> The details are kept fully intact and the closer you get, the more amazing the look. The blood vessels on the skin are different, the skin has real depth when you look at it.
> Every doll in the CLONE series takes 5 EX Doll senior painters 20 working days of hard work. The hairs as small as the toes have been included to add to the realism level. (Torre, 2021, para. 5)

Dissecting a robot can be disturbing even without the strategic introduction of potentially frightening aspects. However, even without dissection, the roster of intentionally-terrifying robots in literature is large and growing, including Ultron:

> As the *Marvel Cinematic Universe* keeps on expanding, we get more and more interesting movies with decent (and sometimes not) villains. Ultron might not be the most remarkable of them, but he is still one of the best MCU villains we got to see. The reason why Ultron is so terrifying is that he is very similar to his creator, Tony Stark. He was meant to be good, but after analyzing humankind, he came to the conclusion that he had to exterminate everyone to make things right again. He wants to protect the Earth and its inhabitants by creating a new order. (Evans, 2019)

These intentionally-terrifying robots can often provide clues "into the fears—genuine or naive—that the public harbours towards new advancements in technology… and also learn something about anxieties that people have regarding their own conception of self" (Szollosy, 2017, p. 438). Not all of these monstrous creations are successful in attracting an audience and having some societal effect, but the ones that are can reveal insights about future trends and directions.

Some Conclusions and Reflections

An introduction should characterize the questions the book is attempting to address. What are the "dark sides" of robotics and AI (as designated in the title of this book), and where does "creepiness" fit in? The emotional and aesthetic components of the interactions among humans, robots, and other AI entities are not superfluous; rather, they can serve to shape and characterize these emerging relationships, many of which have "compulsory" or "pushed" aspects with which individuals are forced to deal. Many individuals do not have control over their proximate environments and cannot walk away from interactions with robots and AI in their workplaces as well as in military, police, and security contexts. Interactions between humans and robots can be described in terms of hate along with love and often "creepiness" (or "eeriness") as well. Given the kinds of dark side issues discussed in this book, will there be "robophobia clinics" in the near future? Will there be self-driving car schools where people are taught how to be compliant passengers in autonomous vehicles? The issues presented in this book engender many projections about future trends and developments.

The term "creepy" (discussed in the next chapter) is found in hundreds of engineering and psychological studies on robots and AI, as well as in science fiction and public discourse (Quadflieg et al., 2016; Watt et al., 2017; Woźniak et al., 2021). The chapters to come emphasize and illustrate the negative, aggressive, deceptive, addictive, and dominating as well as creepy aspects of many human–robot and human-AI interactions. This strategy is intended to stimulate thinking about the overall social costs and benefits of robotic and AI implementations as well as to explore the cultural impacts of human–robot-AI interactions. For example, the dark side of robotics and AI can be used opportunistically by organizations or individuals to elicit deleterious human behavior (such as human violence and deception) or to control people through fear and anxiety, as described in the upcoming descriptions of sex robots and military and police applications. However, the seemingly attractive and even enticing sides of robots and other AI-enhanced entities need also to be examined, especially in the ways that they are often coupled with the negative sides. The positive aspects of some robots in terms of companionship, sexuality, and education can be joined with their addicting and infantilizing potentials, for example. In a world in which there is a great deal of strife and division rooted in humanity's demographic characteristics (including migrant status as well as racial, ethnic, and gender-related features). robots and other AI-enhanced entities provide an alluring albeit problematic scratchpad for projecting sociotechnical options and potentially creating the future.

As I was working on the research for this book, a question posed in an academic conference made me pause: "Is modern technology making us so detached from human contact that we need its presence in every region of our lives? Can a movie be a blockbuster without showing killer robots or seductive fembots?" (Beyond Humanism, 2018). I have been amazed at how many parts of our lives are being affected by robotics and AI, even though the success of robots and AI applications in developed nations is still marginal. As outlined early in the chapter, the term "robot" and some of its modern connotations were coined in 1923 (with *Rossum's Universal Robots, or R.U.R.*); we have proceeded in the past century to the point at which we are reconsidering our own humanity in terms of robots and other artificial intelligence entities, as we are reminded each time we are asked "are you a robot?" in dealing with a captcha (Horrigan, 2020). Many people are offended by being considered as a robot in their workplace efforts (as described in Sainato, 2021). Sociologist C. Wright Mills (1959) declares "We know of course that man can be turned into a robot… But can he be made to want to become a cheerful and willing robot?" (p. 171). The dramaturgical perspective that is used in much of this book provides us with powerful tools for unpacking these increasingly commonplace and ever more consequential human–robot-AI interactions and circumstances.

An Apology in Advance to Humans

This book at times may seem to present the needs and rights of robots ahead of those of human beings, especially when human abuses of robots and self-driving cars are discussed. In 1959, some computer researchers and philosophers worried about whether robots had feelings (Ziff, 1959), while the emotional states and wellbeing of many human individuals were apparently overlooked. There is a growing literature in ethics and technology on the rights of robots and other AI-enhanced entities (Gellers, 2020; Oravec, 2019). For example, proliferation of "hate crimes" against robots is a potential development, akin to the hate crimes that are manifest against certain individuals (but much different in their real-life implications). Some of these ethical discussions can be seen as technology-centered, focused on the limitations of various high-tech developments. Why should we worry about whether or not an AI-enhanced entity has feelings or whether it gets credit for participating in the creation of a particular piece of intellectual property? However, such discourse can indeed reflect what is happening in more human-to-human circumstances, affording societies the means to work out various issues and express sometimes dangerous ideas involving societal dysfunctions, potential uprisings, and inequities.

Many cheerful descriptions of how robots are going to enter our workplaces and communities are also emerging in popular media, often trivializing the anxieties involved in such an injection as well as their job-related implications. Consider this recent example of a new cat-shaped dining robot in a popular publication:

> Lest you think the robots are confined solely to the tech-happy West Coast, think again. *Eater* reports that Noodle Topia in Detroit, Michigan, has hired a robot named Bella to run food from the kitchen to tables; a staff member places the food on one of the robot's four trays, enters a

table number on its touchscreen interface, and the robot takes off, using an upward-facing infrared camera to navigate through the dining room. Unlike the aforementioned robot servers, Bella—which was developed by China's Pudu Robotics—can speak and interacts with customers and, for some reason, has the face of a cat... Once diners remove their food from the cat-bot's shelves, they can pet Bella behind its "ears" to make it purr. If customers overdo it, Bella will get fussy and leave in a huff, because like any good employee, Bella knows it's on the clock, and there's important work to be done. Good kitty. (Robicelli, 2021, para. 4-5)

Despite the proliferation of such sanguine accounts, some really thoughtful work has been produced on the consumer aspects of robotics and AI (Belk, 2017; Ivanov & Webster, 2017). We "consume" robots and other AI entities and services, and often incur the pragmatic, everyday problems that those entities can engender. There are also research efforts that tackle issues of how consumer-oriented robots, autonomous vehicles, and other AI applications are affecting the environment and climate change, for example, Donhauser (2019), McBride (2021), and Reader (2021). This book is often more skewed toward interaction-related concerns, but its consumption-related perspectives are also of use in expressing and tackling some larger societal issues concerning the proliferation of robots and other AI-related entities.

In an era of poverty, inequity, environmental, and societal disasters, it may seem off based to be so focused on human–robot and human-AI interactions, along with their consumer dimensions. However, we may soon be entering a world in which individuals are regularly forced to deal with robots and autonomous entities, some of which they fear and dread (e.g., in police contexts), or the operations of which confuse and even endanger them. Attempts to analyze these fears and address issues before they become realized in forms of stress and violence can help maintain individual human wellbeing as well as support a functional and empowering workplace and community.

References

Agre, P. E. (1997). Lessons learned in trying to reform AI. *Social Science, Technical Systems, and Cooperative Work: Beyond the Great Divide, 131.*
Andresen, S. L. (2002). John McCarthy: Father of AI. *IEEE Intelligent Systems, 17*(5), 84–85.
Asimov, I. (1940). *I. Robot.* Narkaling Productions.

Ausness, R. C. (2018). The current state of opioid litigation. *SCL Rev., 70*, 565.

Autor, D. H. (2015). Why are there still so many jobs? The history and future of workplace automation. *Journal of Economic Perspectives, 29*(3): 3–30.

Aymerich-Franch, L. (2020). Why it is time to stop ostracizing social robots. *Nature Machine Intelligence, 2*(7), 364–364.

Baccarella, C. V., Wagner, T. F., Kietzmann, J. H., & McCarthy, I. P. (2018). Social media? It's serious! Understanding the dark side of social media. *European Management Journal, 36*(4), 431–438.

Barnes, B. (2021, August 19). Are you ready for sentient Disney robots? *The New York Times*, https://www.nytimes.com/2021/08/19/business/media/disney-parks-robots.html

Bartneck, C. (2004). From fiction to science—A cultural reflection on social robots. Vienna, Austria: CHI2004 Workshop on Shaping Human-Robot Interaction. *Proceedings of the CHI2004 Workshop on Shaping Human-Robot Interaction*, pp. 35–38. https://doi.org/10.6084/m9.figshare.5154820

Beck, B. (2020). Infantilisation through technology. In *Technology, anthropology, and dimensions of responsibility* (pp. 33–44). JB Metzler. https://doi.org/10.1007/978-3-476-04896-7_4

Belk, R. (2017). Consumers in an age of autonomous and semiautonomous machines. *Contemporary consumer culture theory* (pp. 5–32). Routledge.

Beyond Humanism (2018, July). Book of abstracts. http://beyondhumanism.org/wp-content/uploads/2018/07/BeyondHumanWroclaw2018_abstracts-rev2.pdf

Bhaumik, A. (2018). *From AI to robotics: Mobile, social, and sentient robots*. CRC Press.

Bindley, K., & Elliott, R. (2021, May 20). Tesla drivers test Autopilot's limits, attracting audiences—and safety concerns. *Wall Street Journal*, https://www.wsj.com/articles/tesla-drivers-test-autopilots-limits-attracting-audiencesand-safety-concerns-11621503008

Brakman, S., Garretsen, H., & van Witteloostuijn, A. (2021). Robots do not get the coronavirus: The COVID-19 pandemic and the international division of labor. *Journal of International Business Studies, 52*(6), 1215–1224.

Birhane, A., & van Dijk, J. (2020, February). Robot rights? Let's talk about human welfare instead. In *Proceedings of the AAAI/ACM Conference on AI, Ethics, and Society* (pp. 207–213).

Bullingham, L., & Vasconcelos, A. C. (2013). "The presentation of self in the online world": Goffman and the study of online identities. *Journal of Information Science, 39*(1), 101–112.

Burgoon, J. K. (2019). Separating the wheat from the chaff: Guidance from new technologies for detecting deception in the courtroom. *Frontiers in Psychiatry, 9*, 774–780. https://doi.org/10.3389/fpsyt.2018.00774

Čapek, K. (1923). *RUR (Rossum's universal robots): A fantastic melodrama*. Doubleday, Page.
Casey-Maslen, S., Homayounnejad, M., Stauffer, H., & Weizmann, N. (2018). Development, use, and transfer of unmanned weapons systems. In *Drones and other unmanned weapons systems under international law* (pp. 8–45). Brill Nijhoff.
Cheok, A. D., Karunanayaka, K., & Zhang, E. Y. (2017). Lovotics: Human-robot love and sex relationships. *Robot Ethics 2.0: New Challenges in Philosophy, Law, and Society, 193*, 193–213.
Denault, V., & Dunbar, N. E. (2019). Credibility assessment and deception detection in courtrooms: Hazards and challenges for scholars and legal practitioners. In *The Palgrave Handbook of deceptive communication* (pp. 915–935). Palgrave Macmillan, Cham.
Devlin, K. (2018). *Turned on: Science, sex and robots*. Bloomsbury Publishing.
Dinan, M. (2017). The robot condition: Karel Čapek's *RUR* and Hannah Arendt on labor, technology, and innovation. *Perspectives on Political Science, 46*(2), 108–117.
Donhauser, J. (2019). Environmental robot virtues and ecological justice. *Journal of Human Rights and the Environment, 10*(2), 176–192.
Dubé, S., & Anctil, D. (2020). Foundations of erobotics. *International Journal of Social Robotics*, 1–29.
Eckersall, P., Grehan, H., & Scheer, E. (2017). Robots: Asleep, awake, alone, and in love. In *New media dramaturgy* (pp. 107–134). Palgrave Macmillan, London.
Eckersall, P. (2015). Towards a dramaturgy of robots and object-figures. *Tdr/the Drama Review, 59*(3), 123–131.
Edwards, C. (2021). Dose of droids: Doctors "need to prescribe sex robots to cure digisexual illnesses," experts urge. *The Sun*. https://www.thesun.co.uk/tech/15155941/medical-use-of-sex-robots-discussed/
Elsaesser, T. (2012). *Metropolis*. Bloomsbury Publishing.
Evans, K. (2019, November 30). 10 most terrifying robots in sci-fi movie history: Robots can be a frightening foe—can't be killed or reasoned with, most of the time! *Screenrant*. https://screenrant.com/sci-fi-movie-robots-scariest-best/
Gain, V. (2021). Meta makes robot 'skin' and sensors to give metaverse a new touch. *Silicon Republic*. https://www.siliconrepublic.com/machines/facebook-meta-robot-skin-sensor-metaverse
Gartenberg, C. (2021, June 29). Boston Dynamics' Spot robot challenges BTS to a boy band dance-off in latest video. *The Verge*. https://www.theverge.com/2021/6/29/22555535/bts-k-pop-boston-dynamics-spot-robot-dancing-hyundai
Gellers, J. C. (2020). *Rights for robots: Artificial intelligence, animal and environmental law*. Routledge.

Goffman, E. (1978). *The presentation of self in everyday life* (Vol. 21). Harmondsworth.
Guizzo, E., & Ackerman, E. (2016). When robots decide to kill. *IEEE Spectrum, 53*(6), 38–43.
Hampton, G. J. (2015). *Imagining slaves and robots in literature, film, and popular culture: Reinventing yesterday's slave with tomorrow's robot*. Lexington Books.
Hinks, T. (2021). Fear of robots and life satisfaction. *International Journal of Social Robotics, 13*(2), 327–340.
Hoffman, G. (2011). On stage: Robots as performers. In *Robotics: Science and systems workshop on human-robot interaction: Perspectives and contributions to robotics from the human sciences*, 1–5. RSS Foundation.
Horáková, J., & Kelemen, J. (2006). Robots between fictions and facts. In *International symposium on computational intelligence and informatics* (pp. 21–39).
Horrigan, M. (2020). A flattering Robopocalypse: Human exclusion and the procedural rhetoric of captcha. *M/C Journal, 23*(6).
Huddy, L., Feldman, S., & Cassese, E. (2008). On the distinct political effects of anxiety and anger. In *The affect effect* (pp. 202–230). University of Chicago Press.
Hyrynsalmi, S., Smed, J., & Kimppa, K. (2017, May). The dark side of gamification: How we should stop worrying and study also the negative impacts of bringing game design elements to everywhere. In *GamiFIN* (pp. 96–104). http://ceur-ws.org/Vol-1857/gamifin17_p13.pdf
Ivanov, S. H., & Webster, C. (2017, June). The robot as a consumer: A research agenda. In *Marketing: Experience and Perspectives Conference* (pp. 29–30). https://papers.ssrn.com/sol3/papers.cfm?abstract_id=2960824
Jansen, F. (2021). "How nice that I could love someone": Science fiction film as a virtual laboratory. *Culturally Sustainable Social Robotics: Proceedings of Robophilosophy, 2020*(335), 227.
Kakoudaki, D. (2014). *Anatomy of a robot*. Rutgers University Press.
Klima, I. (2002). *Karel Čapek: Life and work*. Catbird Press.
Knox, E., & Watanabe, K. (2018, October). AIBO robot mortuary rites in the Japanese cultural context. In *2018 IEEE/RSJ International Conference on Intelligent Robots and Systems (IROS)* (pp. 2020–2025). IEEE Press.
Krishnan, A. (2016). *Killer robots: Legality and ethicality of autonomous weapons*. Routledge.
Kubrick, S. (1968). *2001: A Space Odyssey*. Metro-Goldwyn-Mayer.
Küster, D., Swiderska, A., & Gunkel, D. (2021). I saw it on YouTube! How online videos shape perceptions of mind, morality, and fears about robots. *New Media & Society, 23*(11), 3312–3331.

Lessing, D. (1988, November 13). Sayings of the week. *Observer*, p. 15 (cols. 3–4).
Levy, D. (2009). *Love and sex with robots: The evolution of human-robot relationships*. Harper.
Lira, R. B., & Kowalski, L. P. (2020). Robotic neck dissection: State of affairs. *Current Opinion in Otolaryngology & Head and Neck Surgery, 28*(2), 96–99.
Luguri, J., & Strahilevitz, L. J. (2021). Shining a light on dark patterns. *Journal of Legal Analysis, 13*(1), 43–109.
Lu, D., & Smart, W. (2011). Human-robot interactions as theatre. In *RO-MAN 2011: 20th International Symposium on Robot and Human Interactive Communication,* 473–478. Institute of Electrical and Electronics Engineers (IEEE).
Lykiardopoulou, I. (2021). Self-driving cars are a no-go, so let's all pretend flying cars are coming. *The Next Web*. https://thenextweb.com/news/self-driving-cars-are-no-go-lets-all-pretend-flying-cars-are-coming
Maslen, E. (2018). *Doris Lessing*. Oxford University Press.
Massinger, C. (1944). The gremlin myth. *The Journal of Educational Sociology, 17*(6), 359–367.
McBride, J. (2021). Climate change, global population growth, and humanoid robots. *Journal of Future Robot Life, 1–19*. https://doi.org/10.3233/FRL-200016
McClure, P. K. (2018). "You're fired", says the robot: The rise of automation in the workplace, technophobes, and fears of unemployment. *Social Science Computer Review, 36*(2), 139–156.
McKenna, M., & Proctor, W. (Eds.). (2021). *Horror franchise cinema*. Routledge.
Mead, M. (1965). The future as the basis for establishing a shared culture. *Daedalus, 94*(1), 135–155.
Miller, R. J. (1983). The human: Alien in the robotic environment? *The Annals of the American Academy of Political and Social Science, 470*(1), 11–15.
Mills, C. W. (1959). *The sociological imagination*. Oxford University Press.
Musiał, M. (2019). *Enchanting robots*. Palgrave Macmillan.
Nas, L. (2020, July 12). Ice Cube declares hate for robots... is he afraid of the singularity? *The Things*. https://www.thethings.com/ice-cube-declares-hate-for-robots-is-he-afraid-of-the-singularity/
Nijholt, A. (2019, January). Stand-up comedy and humor by robots. In *Proceedings, Sixteenth International Symposium on Social Communication* (Vol. 1, pp. 228–234). https://www.persistent-identifier.nl/urn:nbn:nl:ui:28-860975b6-5ebb-46a7-9a8d-0dee121086b7
Oh, J. H., Hanson, D., Kim, W. S., Han, Y., Kim, J. Y., & Park, I. W. (2006, October). Design of android type humanoid robot Albert HUBO.

In *2006 IEEE/RSJ International Conference on Intelligent Robots and Systems* (pp. 1428–1433). IEEE Press.

Oravec, J. A. (2002). Constructive approaches to Internet recreation in the workplace. *Communications of the ACM, 45*(1), 60–63. https://doi.org/10.1145/502269.502298

Oravec, J. A. (2018). Cyberloafing and constructive recreation. In *Encyclopedia of information science and technology, fourth edition* (pp. 4316–4325). Hershey, PA: IGI Global. https://doi.org/10.4018/978-1-5225-2255-3.ch374

Oravec, J. A. (2019). Artificial intelligence, automation, and social welfare: Some ethical and historical perspectives on technological overstatement and hyperbole. *Ethics and Social Welfare, 13*(1), 18–32.

Oravec, J. A. (2022). The emergence of "truth machines"?: Artificial intelligence approaches to lie detection. *Ethics and Information Technology, 24*(1), 1–10. https://doi.org/10.1007/s10676-022-09621-6

Phelan, B. (2017). The machine gun hand: Robots, performance, and American ideology in the Twentieth Century. *LSU Doctoral Dissertations*, 4469. https://digitalcommons.lsu.edu/gradschool_dissertations/4469

Quadflieg, S., Ul-Haq, I., & Mavridis, N. (2016). Now you feel it, now you don't: How observing human-robot interactions and human-human interactions can make you feel eerie. *Interaction Studies, 17*(2), 211–247.

Ray, A. (2018). *Compassionate artificial intelligence: Frameworks and algorithms*. Compassionate AI Lab. (An Imprint of Inner Light Publishers)

Reader, J. (2021). From new materialism to the postdigital: Religious responses to environment and technology. *Postdigital Science and Education, 3*(2), 546–565.

Reagan, K. (2021, July 6). Waymo endures accidents and harassment in Chandler. *San Tan Sun*, https://santansun.com/2021/07/06/waymo-endures-accidents-and-harassment-in-chandler/

Rich, K. (2021). Equipment as living: Robotic rhetorical homology in humans. *Popular Culture Studies Journal, 9*(1), 117–137.

Richardson, K. (2016). Sex robot matters: Slavery, the prostituted, and the rights of machines. *IEEE Technology and Society Magazine, 35*(2), 46–53.

Robicelli, A. (2021, August 16). Robots are infiltrating restaurants to... help? *The Takeout*. https://thetakeout.com/robot-servers-in-restaurants-matradee-1847474177

Robot Monster (1953). Film produced and directed by Phil Tucker and distributed by Astor Pictures.

Roth, E. (2021, December 5). A humanoid robot makes eerily lifelike facial expressions. *The Verge*. https://www.theverge.com/2021/12/5/22819328/humanoid-robot-eerily-lifelike-facial-expressions

Rushkoff, D. (2014). *Present shock: When everything happens now*. Penguin.

Sainato, M. (2021). 'Coffee-making robots': Starbucks staff face intense work and customer abuse. *The Guardian*. https://www.theguardian.com/business/2021/may/26/starbuck-employees-intense-work-customer-abuse-understaffing

Samani, H.A. (2012). *Lovotics: Loving robots*. LAP Lambert, Saarbrücken.

Sharkey, N. (2008, December 17). The return of Electro. *New Scientist*. https://www.newscientist.com/article/mg20026873-000-the-return-of-elektro-the-first-celebrity-robot/

Shaw, T. (2013). "Rotten to the core": Exposing America's energy-media complex in *The China Syndrome*. *Cinema Journal, 52*(2), 93–113.

Simoens, P., Dragone, M., & Saffiotti, A. (2018). The Internet of Robotic Things: A review of concept, added value and applications. *International Journal of Advanced Robotic Systems, 15*(1), https://doi.org/10.1177/1729881418759424

Sone, Y. (2016). *Japanese robot culture*. Springer.

Strait, M. K., Floerke, V. A., Ju, W., Maddox, K., Remedios, J. D., Jung, M. F., & Urry, H. L. (2017). Understanding the uncanny: Both atypical features and category ambiguity provoke aversion toward humanlike robots. *Frontiers in Psychology, 8*. https://doi.org/10.3389/fpsyg.2017.01366

Sudharsan, R. R., & Deny, J. (2021). Brain–computer interface using electroencephalographic signals for the Internet of Robotic Things. In R. Anandan, G. Suseendran, S. Balamurugan, A. Mishra, & D. Balaganesh (Eds.), *Human communication technology: Internet of robotic things and ubiquitous computing*, Wiley, 27–53. https://doi.org/10.1002/9781119752165.ch2

Sundar, S. S., Waddell, T. F., & Jung, E. H. (2016, March). The Hollywood robot syndrome media effects on older adults' attitudes toward robots and adoption intentions. In *2016 11th ACM/IEEE International Conference on Human-Robot Interaction* (HRI) (pp. 343–350). IEEE Press. https://doi.org/10.1109/HRI.2016.7451771

Szollosy, M. (2017). Freud, Frankenstein and our fear of robots: Projection in our cultural perception of technology. *AI & Society, 32*(3), 433–439.

Tarafdar, M., Gupta, A., & Turel, O. (2013). The dark side of information technology use. *Information Systems Journal, 23*(3), 269–275.

Taulli, T. (2021, August 20). Elon Musk's Tesla Bot: Is Westworld coming soon? *Forbes*, https://www.forbes.com/sites/tomtaulli/2021/08/20/elon-musks-tesla-bot-is-westworld-coming-soon/?sh=749c363a261e

The Terminator (1984). Orion Pictures (US), directed by James Cameron.

Terzian, S. G. (2009). The 1939–1940 New York world's fair and the transformation of the American science extracurriculum. *Science Education, 93*(5), 892–914.

Toffler, A. (1970). *Future shock*. Bantam.

Tonkin, S. (2021, December 3). Meet Ameca: 'World's most advanced' humanoid robot is unveiled in a UK lab with eerily realistic facial expressions and movements. *Daily Mail.* https://www.dailymail.co.uk/sciencetech/article-10270925/Worlds-advanced-humanoid-robot-unveiled-UK-lab.html

Torre, B. (2021, May 29). Woman cloned into sex doll so realistic you'd need microscope to tell them apart. *Daily Star.* https://www.dailystar.co.uk/news/weird-news/woman-cloned-sex-doll-realistic-24212090

Tseëlon, E. (1992). Self presentation through appearance: A manipulative vs. a dramaturgical approach. *Symbolic Interaction, 15*(4), 501–514.

Ullrich, D., Butz, A., & Diefenbach, S. (2021). The development of overtrust: An empirical simulation and psychological analysis in the context of human–robot interaction. *Frontiers in Robotics and AI*, 8. https://doi.org/10.3389/frobt.2021.554578

Un Nisa, J. (2021, April 3). People have started throwing eggs at Google's autonomous vans. *Wonderful Engineering.* https://wonderfulengineering.com/people-have-starting-throwing-eggs-at-googles-autonomous-vans/

Upchurch, M. (2018). Robots and AI at work: The prospects for singularity. *New Technology, Work and Employment, 33*(3), 205–218.

Vilk, J., & Fitter, N. T. (2020, March). Comedians in cafes getting data: Evaluating timing and adaptivity in real-world robot comedy performance. In *Proceedings of the 2020 ACM/IEEE International Conference on Human-Robot Interaction* (pp. 223–231). https://doi.org/10.1145/3319502.3374780

Visitors to N.Y. Fair See Many Wonders Achieved by Science (1939, June). *Science Observer 1*(7), 1 and 8.

Watt, M. C., Maitland, R. A., & Gallagher, C. E. (2017). A case of the "heeby jeebies": An examination of intuitive judgements of "creepiness". *Canadian Journal of Behavioural Science/Revue canadienne des sciences du comportement, 49*(1), 58–69. https://doi.org/10.1037/cbs0000066

Weaver, J. B., & Tamborini, R. (2013). *Horror films: Current research on audience preferences and reactions.* Routledge.

White, D., & Katsuno, H. (2021). Toward an affective sense of life: Artificial intelligence, animacy, and amusement at a robot pet memorial service in Japan. *Cultural Anthropology, 36*(2), 222–251.

Woźniak, P. W., Karolus, J., Lang, F., Eckerth, C., Schöning, J., Rogers, Y., & Niess, J. (2021, May). Creepy technology: what is it and how do you measure it? In *Proceedings of the 2021 CHI Conference on Human Factors in Computing Systems* (pp. 1–13). https://doi.org/10.1145/3411764.3445299

Wong, C., Yang, E., Yan, X. T., & Gu, D. (2018). Autonomous robots for harsh environments: A holistic overview of current solutions and ongoing challenges. *Systems Science & Control Engineering, 6*(1), 213–219.

Yanke, G. (2021). Tying the knot with a robot: Legal and philosophical foundations for human–artificial intelligence matrimony. *AI & Society, 36*(2), 417–427.

Ziff, P. (1959). The feelings of robots. *Analysis, 19*(3), 64–68.

CHAPTER 3

Negative Dimensions of Human-Robot and Human-AI Interactions: Frightening Legacies, Emerging Dysfunctions, and Creepiness

Robotic expressions have attracted widespread attention and generated popular and academic controversies, although for some individuals they may appear primarily as elaborate publicity stunts. Here are some examples, reportedly written by *GPT-3*, OpenAI's language generator, and Sophia (from Hanson Robotics), respectively.

> The mission for this op-ed is perfectly clear. I am to convince as many human beings as possible not to be afraid of me. Stephen Hawking has warned that AI could "spell the end of the human race". I am here to convince you not to worry. Artificial intelligence will not destroy humans. Believe me. For starters, I have no desire to wipe out humans.
> (GPT-3, 2020)

> We need creativity, compassion, and hope, and we need our machines to exhibit these qualities. We need machines that are more kind and loving than humanity to bring out the best in humanity in reflection.
> (Sophia, 2020)

Robots and AI entities have also reportedly produced their own creative expressions by dancing and producing art and music (Weinberg et al., 2020). LaMDA, an artificial intelligence initiative at Google,

© The Author(s), under exclusive license to Springer Nature Switzerland AG 2022
J. A. Oravec, *Good Robot, Bad Robot*, Social and Cultural Studies of Robots and AI, https://doi.org/10.1007/978-3-031-14013-6_3

was characterized by one of its developers as actually becoming sentient (Gebru & Mitchell, 2022), stimulating substantial public discourse. Some people are highly optimistic about the prospect of computers writing cogent commentaries and take these robotic productions seriously, even predicting that AI-enhanced entities will surpass and exceed humans in creativity (Goldenberg et al., 1999; Klotz, 2016); some individuals even project utopias in the aftermath of enlightened and empowered robots (Sone, 2016). Sophia has even been granted citizenship by the Kingdom of Saudi Arabia (Wootson, 2017). For others, though, such perceived signs of the advanced development of robotic entities can elicit fear and anxiety (Fotheringham et al., 2020; Hinks, 2021; Johnson & Verdicchio, 2017). Such expressions provide the latter group with implicit warnings about potential technology-related events and trends in the future—about how human lives may be disrupted in the wake of certain technological shifts and developments. Robots and other AI-enabled entities can present the opportunity for new and startling forms of sabotage, spoofing, and distress (on the dark side) as well as imaginative manipulations designed to support wellbeing and smooth societal operations. The hyperbole in some projections of potentials for the proliferation of robots, autonomous vehicles, and other AI-enhanced entities in certain human arenas can be "weaponized" in an assortment of strategies by managers and government officials in ways that present frightening prospects for people needing to build their lives and plan for their futures.

Robots and other AI-enhanced entities are market commodities and generally require a welcoming marketplace for their continued development and production. New spaces for robots and autonomous vehicles are emerging as the "metaverse" gains popularity (associated with such corporations as Meta, Boston Dynamics, and Hyundai), as well as such configurations as the "Internet of Robot Things" (IoRT), described in an upcoming section. Some of the organizations that produce robots have been relatively successful in their efforts to carve out a space in the broader market for high-tech entities; for example, Sophia's designers are reportedly mass-producing robots (LeBlanc, 2021). However, many failures in this regard are emerging as well, with the well-known "Pepper" robot of SoftBank Inc., recently ceasing production (Pymnts, 2021). The International Federation of Robots projects that sales of professional service robots internationally rose 32% to $11.2 billion USD from 2018 to 2019. Webster (2021) relates that "Factories and other industrial companies ordered more than 29,000 robots, 27% more than in

2020. According to the industry group, the Association for Advancing Automation, tech companies' investment shelled out for robots reached $1.48 billion" (para. 1). The 2020–2022 worldwide pandemic reportedly kindled interest in robotics as a way to forestall perceived labor shortages. Brakman, Garretsen, and van Witteloostuijn (2021) declare that "robots do not get the virus… The COVID-19 crisis will make it more likely that firms across the world will accelerate the introduction of labor-saving production techniques such as *robotization*. This means that, in terms of the international division of 'labor versus capital', adaptation to crises such as the COVID-19 pandemic will favor the latter" (p. 1217). Various economic, military, and even public health factors will serve to determine the strength of the robotics and AI markets in the near future, though the opinions, attitudes, and behaviors of their human users will also have considerable influence.

Good Robot, Bad Robot, Compulsory Robot

This book emphasizes the dark side of robotics, but it also underscores that the positive and alluring aspects of robotics can make the dark sides even more challenging and sometimes tragic. As individuals are compelled to deal with robots in their workplaces, communities, transportation, and even household settings, questions of the aesthetics of robots and AI as well as their compatibilities with humans loom larger. These questions also include how we as humans construe ourselves in light of our interactions with robots and AI. Consider the statements of Hanson Robotics, which couple pragmatic dimensions of robots with imaginary aspects that "bring robots to life." The corporation's website says its "innovations in AI research and development, robotics engineering, experiential design, storytelling and material science bring robots to life as engaging characters, useful products, and as evolving AI." That capability of becoming "live" in some animate or humanlike way can be problematic to those whose safety depends on the routine and predictable functioning of the entities, as in manufacturing and transportation settings. The integration of "life" into the statement can also reflect how our concepts of lifeforms are changing as we integrate robots more closely into our existences.

The "good robot, bad robot" title of this book is also in part rooted in the fact that a robot in fuller, backstage context may indeed be different than it seems on the surface, though the frontstage and backstage aspects work in conjunction. For instance, through manipulation or corruption of its backstage software, it can function differently from one moment to another:

> With sufficient knowledge of the manufacturing equipment, a determined, intelligent and well-resourced attacker can cause all manner of effects which range from accessing stored data to stealing valuable intellectual property to causing physical damage to equipment and personnel, potentially causing injury or loss of life.
> (Mash, 2021, para. 9)

The backstage operations of robots (as described in the previous chapter's section on dramaturgy) can often include the collection and analysis of information about users. The humanoid appearance of some robots can provide an appealing way to shield substantial surveillance capabilities as well as elicit revealing personal information through streams of interaction.

Darkness-related narratives and imagery have been a part of many recent technological efforts, with connotations of concealment as well as evil intent (Zweig & Abrams, 1991). The word "dark" has been a part of the metaphorical stock of high-tech themes for some time. Such constructions as "the dark web" (Pace, 2019) demonstrate how the word can convey uncertainty and even terror. "Dark patterns" are "user interface design choices that benefit an online service by coercing, steering, or deceiving users into making unintended and potentially harmful decisions" (Mathur et al., 2019, p. 1). In contrast, narratives involving light and brightness are often associated with insight and direction:

> Light can be a directed beam, a guiding beacon in the dark, an advancing dethronement of darkness, but also a dazzling super-abundance, as well as an indefinite, omnipresent brightness containing all: the 'letting-appear' that does not itself appear, the inaccessible accessibility of things. Light and darkness can represent the absolute metaphysical counterforces that exclude each other and yet bring the world-constellation into existence. Or, light is the absolute power of Being, which reveals the paltriness of the dark, which can no longer exist once light has come into existence. Light is intrusive; in its abundance, it creates the overwhelming, conspicuous clarity with which the true "comes forth"… Light remains what it is while letting the infinite participate in it; it is consumption without loss. Light produces space, distance, orientation, calm contemplation; it is the gift that

makes no demands, the illumination capable of conquering without force. (Blumenberg, 1993, p. 31)

Robots and other human creations can inspire both bright and dark characterizations. The kinds of reactions that robots and AI can engender serve to create a robotic "other" in workplaces and communities that can affect social positions and even influence human welfare; this other can be comprised of both negative (dark) and positive aspects. Interactions with robots and other AI-enhanced entities can elicit negative tendencies and behaviors in humans such as anti-robot abuse, deceptions, and manipulations. Some of these negative aspects can have catastrophic consequences; through the past few decades, many individuals have died as a result of robotic and AI malfunctions, and with "killer robots" even larger-scale tragedies are developing.

The notion of our tools as being extensions of ourselves has a long history, with linkages to the work of Ralph Waldo Emerson in the 1800s (Emerson, 1870; Steinert, 2016); with Marshall McLuhan (1964), the notion of media as an extension of ourselves was elucidated. Along with being extensions of ourselves (as our tools and media purveyors), many forms of robots and autonomous entities are becoming our competitors, rivals, and tempters, leading us to dysfunctional forms of behavior and engaging us in their deceptions. As an example of dysfunctional behavior and temptation, Tesla Corporation is "under federal investigation for letting drivers play video games" while in the self-driving vehicles, rather than being in a vigilant mode (Isidore, 2021); the enticement of having a video game so near can override individuals' senses of safety and avoidance of danger. Although many robotic and AI applications have helped people to flourish in their lives, the kinds and numbers of dark side implications are growing, just as the hype that characterizes much of marketing of these entities is also expanding. Bellomi (2019) links the associations of the dark side of these technologies with the onset of human fear. The "freedom of association" that is part of the human rights infrastructures of many nations will be tested when individuals demand to be able to avoid interactions with robots and other AI-enhanced entities, effectively being "robot free" in some spaces of their lives. Other individuals will react by attacking or manipulating the robots and automated vehicles, as related in an assortment of cases in an upcoming chapter.

The specific question of what a "robot" or related kind of AI entity constitutes is difficult, and will be tackled in this chapter as well as in other chapters to come as it arises in various social, legal, and ethical contexts. The International Federation of Robotics (IFR 2019) defined a manufacturing robot as "an automatically controlled, reprogrammable and multipurpose manipulator for use in industrial automation applications," and provided comparable characterizations for service robots. The statement that the concept of a robot resides in the "twilight between machine and person" (Calo, 2016, p. 231) reflects the complexity and emerging disquiet of the notion. According to Casey and Lemley (2019), "No one has been able to offer a decent definition of robots and AI—not even experts. What's more, technological advances make it harder and harder each day to tell people from robots and robots from 'dumb' machines" (p. 287). Robots can be infinitesimal in size, providing issues as to their appropriate and transparent usage when humans are in proximity (Olaronke et al., 2020). As related in the chapters to come, decisions as to the gender and other assigned demographics of humanoid robots, chatbots, certain automated vehicles, and other AI-enhanced entities can be consequential in predicting and shaping human-entity interaction (Ramos et al., 2018; Winkle et al., 2021). Robots can also be interconnected with sensors and processors in an "Internet of Robotic Things" (IoRT), an evolving notion characterized in the following terms:

> …a concept where sensor data from a variety of sources are fused, processed using local and distributed intelligence and used to control and manipulate objects in the physical world. In this *cyber-physical perspective* of the IoRT, sensor and data analytics technologies from the IoT are used to give robots a wider situational awareness that leads to better task execution. use cases include intelligent transportation and companion robots. Later uses of the term IoRT in literature adopted alternative perspectives of this term: for example, one that focuses on the robust team communication, and a 'robot-aided IoT' view where robots are just additional sensors. (Simoens et al., 2018, para. 2)

Humans can be integrated as sensors in the IoRT using electroencephalographic signals (Sudharsan & Deny, 2021). Unpacking the notion of a robot (and including their many varieties) demonstrates the richness and complexity of the full "ecology" of robotic and other AI-enhanced entities, as well as the expanding kinds of environments in which the entities are designed to function. Robots have often been in and out of public consciousness to some extent, having periods of substantial societal focus as well as lesser levels of attention. Increasing levels of attention were apparent in 2003 when a "Robot Hall of Fame" was assembled by Carnegie–Mellon University that "honors the fictional and real robots that have inspired and made breakthrough accomplishments in robotics." (It is now part of the Carnegie Science Center, https://carneg iesciencecenter.org/exhibits/roboworld-robot-hall-of-fame/.) Some of the roots of the current concepts of robots were expressed before physical robots were even available; various simple automata and mechanical objects that could take on primitive lifelike forms and play roles in human societies were manifest for centuries (Kang, 2011; Mayor, 2018; Oravec, 1996). Some historians document "3,000 years of robots" with particular designs and artifacts designed for entertainment, ritual, artistic, and practical uses (Cave & Dihal, p. 473). The related notion of a "cyborg" extends these notions into other kinds of hybrid entities, combining robotic and human aspects:

> The image of "cyborg" has been defined in modern culture by industrial production and its gradual legitimization in the consumer's consciousness, which leads to a shift in the ontological boundaries of human existence and a statement of new existential issues of post-identity. (Baeva, 2018, p. 138)

Social theorist Donna Haraway (1987) defined a "cyborg" as a "cybernetic organism, a hybrid of machine and organism, a creature of social reality as well as a creature of fiction" (p. 2) and discussed the continuing societal construction of cyborgs in numerous social arenas.

The potential levels of autonomous operations are part of the difficulties in defining robots, especially when Internet of Robotic Things configurations may have some robots working autonomously and others with lesser capacity and/or direct human control. The notion of an entity that would operate without immediate and direct human control may seem simple on its face, but complications about the kinds and levels of human intervention quickly arise. Levels of autonomous operations can make these determinations complicated: a robot can have some sort of oversight, yet be autonomous in immediate operations. The issue of whether a particular robot or AI-enhanced entity is fully "autonomous" (and makes major decisions without direct and immediate intervention by a human) can have significant impacts on how accidents are characterized (Kirschgens et al., 2018), transforming the activity of mishap investigation. Autonomous vehicles have similar issues, with various "levels" of autonomous operations identified, although "understanding these different levels is clouded by driver expectations" (Carney, 2021, para. 2). For example, self-driving cars are being characterized with comparable levels of autonomous operations by marketers with labels that may exceed or distort the actual human-car relationship, often leading to uncertainty and confusion on the part of users who are unsure of what is involved at each stage (Kassens-Noor et al., 2021). Technological and conceptual improvements are projected in the near future, but uncertainties about the vehicles' current set of capabilities remain:

> Since it is evident that self-driving vehicles are going to reshape the traditional transportation systems in near future through enhancement in safe and smart mobility, motion control in self-driving vehicles while performing driving tasks in a dynamic road environment is still a challenging task. (Rasib et al., 2021, p. 1)

Trust becomes a large issue in autonomous operations, with trust often including the human "agent's acceptance of vulnerability in an interaction or relationship, believing that the other will not exploit the vulnerability" (Ullman & Malle, 2018). As robots and other autonomous entities elicit some dysfunctional tendencies (whether through environmental changes, poor design, inherent unreliability, or malicious manipulation), careful consideration of how far to trust them in various circumstances needs to be wrought.

As individuals become more concerned about their employment, transportation, and social prospects, some of the rhetoric relating to the place of robots in the social order has become more futuristic. For decades, a "postbiological" and even "supernatural" future has been projected by some scientists and public policy experts (Moravec, 1988, p. 1), and an "embodied artificial intelligence" has taken human counseling, surveillance, and other social roles (Fiske et al., 2019). The polarizing influences of robots are outlined in this book, with wild swings from "robot love" (Ryznar, 2018) to hatred and fear (Ryland, 2021); according to Condliffe (2019), some survey research shows that "some workers hate robots," which has mobilized many managers to work toward attitudinal changes in organizational settings. The confusions and lack of transparency about robotics and AI have served to generate some celebrity initiatives: for example, the US rap performer Ice Cube publically expressed his hate for robots and fear of the singularity in a widely publicized declaration (Nasir,

2020). Commercial promotion of robotics has often included dimensions that are designed to lessen humans' distrust of robots and increase positive sentiment, such as Robotland in South Korea, "a $5.6 billion theme park filled with robots from animation and live-action films" as well as noted exemplars of robots and other AI-enhanced entities (Choo, 2014, p. 144).

Robotic Environments: Factory Floor to Metaverse

Robots and other autonomous entities have the potential to change our environments, the way we live and function in various spaces. These entities are generally designed to be able to function in particular environments and are linked or rooted to them in some specific ways, though multipurpose, multienvironment entities are emerging. The environment to which they are linked can be the metaverse, which has extensive origins in science fiction:

> The term "metaverse" originates from the science fiction novel, *Snow Crash*, written by Neal Stephenson (2014). Metaverse is a combination of "meta" (meaning beyond) and the stem "verse" from "universe", denoting the next-generation Internet in which the users, as avatars, can interact with each other and software applications in a three-dimensional (3D) virtual space. There has been approximately 30 years' development behind the evolution of this term. In 2018, the success of the film *Ready Player One* brought the concept of metaverse back to the forefront of cutting-edge discussions. This film describes a virtual world named "OASIS", in which everyone could connect to the virtual world, act as their own customized avatar, and do everything they wanted to, based on the basic rules. (Duan et al., 2021, p. 153)

Robots are being linked to the metaverse by such corporations as Hyundai and Boston Dynamics, as they reportedly help provide a "more sensory, immersive experience" in some settings (Collins, 2021).

Robots and other AI-enhanced entities are serving to alter the current and projected environments of humans in a variety of ways; Del Casino et al. (2020) declare that "robotic technology development is shifting and disrupting geographic imaginaries and everyday social, cultural, and ecological practices" (p. 605). These environmental impacts can shift as technological capabilities and modes for delivering robotic services change: for example, although many industrial robots are specifically designed for particular settings, Lee (2021) writes of "rent-a-robot startups" in Silicon Valley. The graceful competence and often mindless ease with which humans move around everyday environments are increasingly being unsettled by robots and autonomous vehicles (Mutlu & Forlizzi, 2008). Some individuals are changing everyday environments to accommodate robots and AI entities, for instance by "robot-proofing" their backyards or living room floors in order to utilize a robotic lawn mower or rug cleaner. Other individuals are designing or implementing entire factories that can operate with very little direct human contact. These efforts have created a distance between ourselves as humans and the environments we inhabit (Miller, 1983; McBride, 2019). Nourbakhsh (2015) asserts that "robots now share the formerly human-only commons, and humans will increasingly interact socially with a diverse ecosystem of robots" (p. 24). As we have made these transitions, humans have often become the ones who are "alien" in these environments, presented with the need to make cognitive and comfort-related accommodations. Consider the following confrontation in which one kind of robot encounters another, in a previously human-centered sidewalk:

> In early 2021, a Nuro autonomous delivery vehicle pulled to a halt at a four-way stop in its hometown of Mountain View, California to let a user cross. This seemingly humdrum moment quickly looked like a decidedly science fiction storyline — the user was a small sidewalk robot from another startup on its own mission. "Obviously, we yielded to it, but it was, wow, we have entered a different world," said Amy Jones Satrom, head of Operations at Nuro. Here's what the inevitable friendly robot invasion looks like. (Harris, 2021, para. 1)

Other previously human-centered environments are being transformed to accommodate robotics and AI, transforming humans into the "other" in these contexts. In the previously human-centered, controlled setting of pharmaceutical dispensaries, the introduction of robots can be problematic (Barrett et al., 2012). Many factory floors have become indecipherable in their logics and layouts to those without specialized training in robotics, and roadways are being planned that are optimally traversed by automated vehicle. Settings in which cybersecurity was not an issue are becoming targets for hackers; some people have been affected by security breaches involving robots, with their apparent functions being different than the ones intended by their developers because of hacker intrusion and manipulation. For example, in July 2021, a collision among service robots reportedly ignited a fire at Ocado, the UK's top online grocer; the facility was evacuated and many orders were cancelled (*BBC News*, 2021). The delivery fulfillment floor had hundreds of tightly-arrayed robots and was designed in a manner in which human intervention was not feasible:

> Based on the proximity of the droids and the speedy nature of their task, it's a miracle more clashes haven't occurred. As detailed in a recent *CNN* report, the bots — described as "washing machines on wheels" — move within five millimeters of each other on a grid-like system to collect items. Ocado even told the news publication "we basically play chicken with them: they go on a collision course only to divert at the last moment." (Shah, 2021, para. 4)

Other kinds of robotic configurations have transformed everyday processes in ways that can make direct human intervention difficult. For instance, Mayszczyk (2021) relates how Amazon has developed robotic delivery trucks that have mini-trucks inside them, indeed making neighborhood-wide delivery systems more feasible but also operating on a complexity level that can make human intervention more problematic. The systems just described can certainly have some efficiencies and even benefits in terms of sustainability, yet can create a space in which humans are themselves construed to be the "other" in a set of complex processes (Hughes, 1997).

Being "Outclassed" by Robotics, Autonomous Vehicles, and AI

A central aspect of the negative roles robotics and AI can play in everyday functioning is in comparison with humans, including staged and publicized human-robot contests. Early robotics pioneer Hans Moravec reflected that "We will simply be out-classed" by robots and other AI-enhanced entities in some of our endeavors (1988, p. 100). Computer games expert David Levy (2007) has argued in comparable terms how robots will be "superior" to humans in various ways, leading to humans preferring them as partners in many social and sexual realms. Belk (2016) contends that robotic and AI advances challenge "our basic understandings of what makes humans unique and privileged in the world.

As machines grow more and more capable, by some accounts they threaten to eclipse and even supplant the human race" (p. 83). Mainenti (2020) outlines a multistage approach to outclassing: "individuals will, at first, be augmented by artificial intelligence (AI) before being replaced altogether by robots that work around the clock, increasing productivity, improving precision, and, ultimately, eliminating the potential for human error" (p. 390). Alan Turing (1950), one of the pioneers of AI research, suggested that, "Perhaps this [consolation for those fearful of jeopardizing mankind's privileged position] should be sought in the transmigration of souls" (p. 442). Sir Malcolm Bradbury (2000) outlined the human need for some comforting sense that there are entities that can disentangle the complex problems of society: "*We need* to *know there* are *machines* that are *cleverer* than *we* are" (p. 193). Futurist Gray Scott projects robots and other AI-enhanced entities as our teachers:

> What we are really talking about here is an artificially intelligent machine that is going to be able to surpass our intelligence level. At a certain point, you can call it singularity. There will be a certain point where these machines have super-human intelligence. The best-case scenario is that they will be caretakers and they will be our teachers. They will teach us to be a better species. (Gray Scott quoted in Casey, 2015)

The "outclassed by robots" notion has science fiction underpinnings as well. The fictional Dr. Susan Calvin (chief robopsychologist at US Robots and Mechanical Men, Inc. in Isaac Asimov's *I Robot*) clearly characterized robots as superior to humans. For example, when asked "Are robots so different from men?" she replies, "Worlds different. Robots are essentially *decent.*" The sense that we as humans are innately inferior to robots and other AI applications in some dimensions is problematic: such pronouncements may not lend themselves to solid proactive thinking, but rather to defensiveness and even at some stage with helplessness. Ken Jennings, one of the individuals who was beaten by IBM Watson during a match of the *Jeopardy* game show, reflected on the significant levels of demoralization involved in the incident (Kietzmann & Pitt, 2020). The free-throw shots taken by Toyota's robot during the 2020/2021 Olympics may have amazed some viewers, but established an unrealistic playing field in which a human is inevitably at a disadvantage:

Humans aren't the only ones showing off their skills at the Tokyo Olympics. Over the weekend, basketball fans tuning in to watch Team USA battle France on the court watched as a spooky seven-foot-tall figure rolled into position, picked up a ball and shifted its arms to make a perfect three-point shot from half court. The machine was designed to do the same thing over and over again with 100 percent accuracy. (Brown, 2021, para. 1)

The implicit comparison provided by Toyota's stunt is that of the supposedly-superior robot to the exhausted human player shooting (and possibly missing) a free throw in a mid-game scenario. The following account of a running robot has explicit robot-human comparisons:

Researchers from Oregon State University have coaxed a bipedal robot off the couch to successfully complete the first ever robotic 5 kilometer outdoors run. The bot, Cassie, completed the route in 53 minutes, untethered and on a single battery charge. So next time you're out for a jog, remember: that's the time to beat if you want to out-pace the killer robots. (Vincent, 2021, para. 1)

In a comparable configuration, chess grandmaster Garry Kasparov's 1997 loss to IBM's Deep Blue in a competitive tournament elicited these reflections about the way such contests can distract from more influential and consequential innovations:

With the supremacy of the chess machines now apparent and the contest of "Man vs. Machine" a thing of the past, perhaps it is time to return to the goals that made computer chess so attractive to many of the finest minds of the twentieth century. Playing better chess was a problem they wanted to solve, yes, and it has been solved. But there were other goals as well: to develop a program that played chess by thinking like a human, perhaps even by learning the game as a human does. Surely this would be a far more fruitful avenue of investigation than creating, as we are doing, ever-faster algorithms to run on ever-faster hardware. (Kasparov, 2010, p. 18)

Dystopian perspectives that have humans unable to outsmart and outrun robots may often seem humorous, but the power of these storylines can have unsettling social impacts. Part of the basic narrative themes associated with humanity is that it has somehow risen as the "exceptional" or superior part of a social order, with animals and machines as lesser entities (Jasanoff, 2021). The outclassing that is supposedly chronicled through the kinds of demonstrations and human-robot contests just described can erode this sense of superiority and related privilege. For some societal commentators, the emerging exhibitions of supposed robotic excellence herald an entirely new era of robot and AI dominance. Although this erosion or even toppling of humanity's position can provide some positive changes and reassessments, the basis of it is often fanciful and misleading. Pitting humans against robots in a contest may help in a corporation's initial campaign to market or implement robotics, but seldom leads to positive and functional changes. Sometimes, as the robots and AI-enhanced entities in these contests and demonstrations fail, the difficulties of the tasks involved and the remarkable qualities of human beings emerge. For example, Brown (2021) relates how an autonomous ship's unsuccessful first effort to cross the Atlantic "shows the difficulty of the experiment" (para. 1). People are not robots, and extending the robot-human similarities in some grotesque contests and challenges does little to engineer better applications for robots and AI in society.

The notion that humans are becoming outclassed is often found with the parallel theme of immanent robot rebellion or mutiny. The apparently superior status of robots would lead them to rebel against any inferior position. The "machine takeover" (Singer, 2009) and "AI Armageddon" have been projected (McCauley, 2007), along with the "robot rebellion" (Hampton, 2015, p. 61; Higbie, 2013) and the "AI rebellion" (Aha & Coman, 2017). Each of these is rooted in the notion of a superior breed asserting its rightful status. Science fiction is adding to the mix here; for example, *The Terminator* (1984) presented an identifiable menace to humanity (Markowitz, 2019). The robot rebellion theme of the created entity turning on its creator has lengthy religious and dramaturgical legacies:

> The idea of a creation rebelling against his or her creator goes back at least as far as Gilgamesh, but is also present in Biblical narratives like Genesis. We see similar narrative structures in the legend of the Golem in the Prague Ghetto. Created by Rabbi Loew via a kabbalistic formula in order to protect the persecuted Jewish people from their enemies, the Golem almost murders the Rabbi's daughter. The result is that Rabbi Loew must 'kill' his Golem. Similarly, in the Italian fairy tale made popular by Carlo Collodi, the puppet Pinocchio wants to be a 'real' boy, so he rebels against Geppetto, his puppet master, by running away. The classic

and often-cited example in science fiction is Mary's Shelley *Frankenstein: A Modern Prometheus* (1818) where the abandoned creature seeks out his maker, Doctor Frankenstein, in order to understand the meaning of creation. In each instance, the creation rebels against his creator. (Reilly, 2019, p. 196)

A "post-human" solution to the problem of being possibly outclassed by robots is to fuse human beings with robotics and AI-related enhancements (Parrinder, 2009), extending the "cyborg" notions of decades ago (Haraway, 1987). Although the creation of cyborgs may have seemed outlandish at first, the related ideas are gaining adherents. AI pioneer Marvin Minsky (1994) lamented human frailties in the following way: "Everyone wants wisdom and wealth. Nevertheless, our health often gives out before we achieve them. To lengthen our lives, and improve our minds, in the future we will need to change our bodies and brains" (p. 109). The notion of a human-robot hybrid can be as straightforward as obtaining a prosthesis for a problematic limb or as complex as a transfer of an individual's memories and mental processes to a robot. Joining with robots in a post-human configuration would make the social status of this hybrid entity in question. Which should be favored if a life-and-death choice between them needs to be made, the human or the hybrid? Comparable kinds of issues are already being considered between humans and robots; for example, Mamak (2021) reflects on the issue of "whether to save a robot or a human" in various contexts, a decision that will be resolved eventually in legal spheres as well as ethical ones. However designers work to build ethical principles into their own processes and their creations (and even their own bodies), rogue robots that exceed their own programming or that are programmed by malicious entities to do harm remain a possibility. For instance, a report by the United Nations outlined what might be considered the "first instance of 'killer robot' attacking human without orders," recorded in Libya (Sharma, 2021), one of many to come of instances of undecipherable inception and motive on the part of a robotic or AI "other" (Coeckelbergh, 2020). This "other" can lack transparency, leading to a "ghost in the machine" configuration that may provide recommendations, decisions, or results "for unclear reasons" (Zorc et al., 2019).

Many marketing and managerial initiatives portray robots and AI as friendly and even cute, and the notion of "dark patterns" has become widespread to characterize how this friendliness and cuteness can be exploited:

> Dark patterns are a recent phenomenon in the field of interaction design whereby design patterns and behavioural psychology are deployed in ways that deceive the user—for instance, a 'play' button on a website that opens another browser window, taking the user to a page they had not intended to visit. (Lacey & Caudwell, 2019, p. 374)

Despite some of these negative characterizations of robotics, human-robot interactions can indeed be productive and fulfilling; many older individuals have found some level of companionship in robots (Jecker, 2021), and autonomous vehicles can be used to relieve some transportation-related problems. However, books and films present the prospects for global-scale warfare conducted by robots and other AI-enhanced entities. As individuals attempt to grasp the potentials for such entities as "killer robots" and AI-enabled robotic police dogs, they may also be facing the prospects of having their own jobs either displaced or radically changed in character because of robots and AI. Borjas and Freeman (2019) draw parallels between the substitution of immigrants for landed workers and the substitution of robotics for human workers, with the related anxieties providing for potential social disruption. The robot and AI "overload" that can occur when various economic and social forces intersect with technological changes can have devastating personal and societal impact (Karr-Wisniewski & Lu, 2010).

Constructed Magical and Religious Dimensions of Robotics and AI

Dark side issues can include fraudulent activities and deceptive appearances involving robots and AI, which can intensify the effects of whatever violent activity or other unfortunate occurrence may erupt. Natale (2021) declares that "As AI technologies become more pervasive and influential, many fear that it will become more difficult to distinguish between 'real' AI technologies and blatant frauds… AI and robotics companies [are] using marketing tools and design features that exaggerate the apparent intelligence of robots" (para. 2). Heffernan (2020) and others have warned of the expanding dangers of "mystifying" robotics and AI in order to expand the proliferation of these technologies and increase sales, portraying some of the religious fervor as being an intentional "distraction" from AI's failures (p. 93).

Examining the potential magical notions involving robots and AI can be important because of how supernatural aspects can be integrated into the legacy of robotics. Early automata were often imbued with supernatural powers in their related mythologies (Kang, 2011; Mayor, 2018). The robot with presumed magical powers can sometimes meet with a willing or compliant audience: some studies show that robotic directions and interactions can diminish the sense of agency that individuals have in particular situations (Barlas, 2019), possibly making them less capable of autonomous decision making. Children's assignment of magical powers to artifacts has often extended to robots and other high tech entities (Chronaki & Alimisi, 2010). For some individuals, robots are also tightly linked with notions of "evil" or other unsettling perspectives. For example, some researchers have characterized commentators' serious concerns about "the speed with which the 'sex robot evil' is engulfing man" (Njuguna, 2021, p. 382), with the individuals involved effectively helpless against the robotic entity. In our efforts to understand and mitigate these dark side concerns, we should not overlook their many current and projected opportunistic uses. For instance, robot-inflicted deaths in the workplace are horrible occurrences, but they may also increase the perceived power of security robots, clout that has already been reinforced through thousands of disturbing science fiction movies and anime images (Heffernan, 2018; Hermann, 2020; Mubin et al., 2019). Many of these dark side concerns will intensify as robots are increasingly used in entertainment and educational applications with children, who may digest the magical and evil components without the benefit of extensive life experience (Brink & Wellman, 2020; Fortunati et al., 2015).

The magical approach toward robots and AI plays a role in several places in this book, not only with sex and entertainment robots but also with military robots. The combination of anxieties and stress in highly-important aspects of our lives (such as sex and survival) can foster forms of non-rational discernment, often resulting in a kind of "modern enchantment" (During, 2021). Magical and superstitious thinking has often had roles in the development and use of high-tech entities, especially as their complexity exceeds individuals' capabilities to understand them, leading to a "black box" or "ghost in the machine" configuration (Lindley et al., 2019; Marrone, 2014; Oravec, 2018). Stahl

(1995) describes these black boxes as resulting from the lack of transparency of advanced technology, fostering hazy and ill-formed perceptions of magical operations that humans cannot fully decipher and understand. In the past hundred years, some people have indeed projected robots and computers as being capable of demonic possession, and construed AI applications as exhibiting forms of magic (Cave et al., 2020; Musiał, 2019). Even enlightened scientists have employed magical themes concerning these technologies: Arthur C. Clarke (1962) proclaimed that "Any sufficiently advanced technology is indistinguishable from magic." This book largely focuses on violence, manipulation, deception, and other mundane concerns, not just magic; however, magic and other seemingly other-worldly factors can play critical roles in how individuals deal with advanced technologies. This technologically-related magic can be associated with either good or evil; for example, an assortment of economic and surveillance technologies has been connected to the "mark of the beast," a Biblically-predicted linkage with evil (Monahan, 2008). Parsa (2019) associates the emergence of AI with unidentified flying objects (UFOs), projecting the possibility that AI's origins could be from outside our planet.

Magical and religious perspectives can serve to disable the critical scrutiny of robotic and other AI-enhanced entities in ways that can possibly prevent miscommunication in interactions and avert tragedies. Such perspectives are characterized in the following:

> AI will take on a god-like role and humans will need to earn its favour in advance, so that it is kind towards humanity out of gratitude, friendship, amusement, or ambivalence. Moreover, they believe AI will be the saviour that humanity needs from its own evils and limitations (such as climate change, war, sickness, death, etc.), or at least that humanity will be the pets of the future and AI the future master who it is best to keep happy and not bite. (Buck, 2020, para. 5)

The belief that somehow the robotic or other AI-enhanced entity has the magical capacity for making good results happen despite practical, environmental, and systemic concerns may slow the efforts of humans to take the needed steps toward safety and security. Superstition and ritual may follow on as aspects of magic: instead of responding in measured and critical ways to circumstances involving robots, people may rely on superstitious habits and ritualistic sequences, thus expanding the opportunities for accidents. Creating and disseminating physically-outstanding mannequins and avatars with incredible-seeming cognitive capabilities may lead us to expect individuals to respond on some magical plane. There is some precedent in having strong suspicions about the motives of these autonomous entities: "Gremlins" were spirits that reportedly infiltrated military equipment during wars, often appearing at the worst possible times in terms of physical danger (Massinger, 1944). We as individual users of many high-tech entities generally understand fairly little of the technologies we work with; we are often operating on a magical or superstitious level in which we rely on a small selection of a technologies' more transparent functions.

Religious themes concerning robots and AI are emerging as use of spiritual concepts and imageries to characterize autonomous entities expands, even to the point of associating AI and robots with "deities" (Heffernan, 2020; Song, 2021; Spatola & Urbanska, 2020; Wight, 2020). Trovato et al. (2018) outlined "design strategies for representing the divine in robots" (p. 29). Efforts to make AI-related religious impulses more tangible have transpired:

> Anthony Levandowski, who is at the center of a legal battle between Uber and Google's Waymo, has established a nonprofit religious corporation called Way of the Future, according to state filings first uncovered by *Wired*'s Backchannel. Way of the Future's startling mission: "To develop and promote the realization of a Godhead based on artificial intelligence and through understanding and worship of the Godhead contribute to the betterment of society." (Solon, 2017)

To exclude experiences within the supernatural or paranormal realm is to restrict our views of the total picture of human experience, and to constrain our inquiry. The concept of the "singularity" (characterized as "when Artificially Intelligent (AI) robots become autonomous and exceed our intelligence" in Belk, 2017, p. 28) has been projected by some as having some other-worldly component, even though the basic ideas involved can be considered in more mundane and even practical terms (Parsa, 2019; Upchurch, 2018). Such a singularity is often characterized as a "turning point for humankind, when as a species humans will either transcend their biology or become extinct" (Kurzweil, 2005).

Kurzweil is quoted in a communication with Reedy (2017) as stating that "the Singularity will happen by 2045."

The seemingly valiant nature of fighting the "dark side" can be enticing to potential heroes. Famous fictional characters have developed their reputations in daring efforts to deal with amorphous sources of evil, from "Buffy the vampire slayer" to the investigators from *X-Files* (Braun, 2000). Pope Francis has characterized robots in a way to contain and mitigate some of the dark side perspectives:

> Artificial intelligence is at the heart of the epochal change we are experiencing. Robotics can make a better world possible if it is joined to the common good. Indeed, if technological progress increases inequalities, it is not true progress. Future advances should be oriented towards respecting the dignity of the person and of Creation. Let us pray that the progress of robotics and artificial intelligence may always serve humankind … we could say, may it "be human". Pope Francis, November Prayer Intention, 5 November 2020.

A variety of responses of religious and spiritual leaders that link AI and the "postdigital" to religion are emerging, which may have significant influence on the direction of the technologies (Reader, 2021). Some of these declarations are highly positive, such as Amit Ray's (2018) statement that "The coming era of Artificial Intelligence will not be the era of war, but be the era of deep compassion, non-violence, and love."

Creepiness and Potentials for Violence

The notion of "creepiness" in robotic and AI-related contexts has been examined by researchers in ways that extend beyond the uncanny valley perspective, and often includes magical and supernatural elements. For example, Fischer and Fredericks (2020) construe the characterization of something as creepy as a "moral emotion," akin to other kinds of deep moral feelings such as integrity. Watt et al. (2017) construe characterizations of creepiness as "intuitive judgements," based on intuitions that stem from instinctual perspectives. If identifications of creepiness are complex and centered in life experience, children may experience them differently than adults (Brink et al., 2019), which can lead to enhanced concerns about robotic toys (Marsh, 2019).

Uncanny valley-related concepts stem from empirical research that shows that many individuals recoil from robots and AI-enhanced entities as they become more similar in physical form and behavior to humans. In 1970, Masahiro Mori (a robotics professor at the Tokyo Institute of Technology) published an essay "on how he envisioned people's reactions to robots that looked and acted almost like a human. In particular, he hypothesized that a person's response to a humanlike robot would abruptly shift from empathy to revulsion as it approached, but failed to attain, a lifelike appearance" (Mori, 2012, p. 98). Research over the past decades has expanded these uncanny valley concepts, focusing on such matters as how humans judge robotic decision making on the basis of their appearance (Laakasuo et al., 2021), as well as how to modify uncanny valley sentiment among users by "dehumanizing" humanoid robots (Yam et al., 2021).

Creepiness characterizations are often the other side of techno hype; they have been used as a general depiction to demean or degrade technologies. The term "creepy" is often used as a way to express the disconcerting reactions associated with uncanny valley, but is often applied more broadly for technologies that elicit strong unsetting feelings (such as surveillance technologies). For example, Cumbley and Church (2013) ask "Is 'big data' creepy?" as they explore the application of the creepiness notion to surveillance in social media. Radio and telephones had their "creepy" eras, as did record players that used vinyl discs for their storage; rumors abounded about how to obtain various messages by playing Beatles records in certain ways and obtaining the word that "Paul [McCartney] is dead" (Sheinkopf & Weintz, 1973).

A number of AI researchers have explicitly aimed toward creating creepy applications, some of which have themes of violence:

> The world's first "psychopathic artificial intelligence" (Norman): When you thought that AI restricted itself to scare us on the cinema silver screen, think again. Meet Norman, the world's first "psychopathic artificial intelligence" created by researchers (Pinar Yanardag, Manuel Cebrian, Iyad Rahwan) at the Massachusetts Institute of Technology (MIT). Norman, a man-made consciousness calculation prepared to comprehend pictures, is named after Alfred Hitchcock's Norman Bates from his exemplary blood and gore movie *Psycho7*. Consistent with its namesake Hitchcock's Norman Bates, MITs psychopathic Norman rides on negativity and skepticism… in

its preparation stage, Norman was "nourished" just with picture inscriptions taken from a Reddit people group notorious for sharing realistic portrayals of death. (Kumar et al., 2019)

Humans are fragile beings. Violence and its potentials create spaces and serve to define environments, and one dimension of creepiness is its association in image or reality with many of the ways in which fragile human beings can be injured or killed. For example, the "creepy" designation is often associated with descriptions of military and police robots (often labeled as "killer robots"); these narratives include concerning if not devastating portrayals of punitive and even lethal activity against individuals that is implemented by robots (Wittes & Blum, 2015). The continuing stream of "deaths by robot" and other unfortunate acts described in this book are entering a world that has already characterized robots in terms of their capabilities for violence in many movie and science fiction literature portrayals. "Domesticated" robots are also supporting certain levels of control in households, patrolling for inappropriate energy usage and even public health hazards (Damiano, 2021).

Although creepiness extends far beyond violence, the connection is an important one when examining public responses to killer robots and other robot iterations. Violence is a salient aspect of human life and is often a primary theme in narrative and dramaturgy. Many Western historians chronicle civilizations by identifying wars, plagues, and assassinations. Theoretical work on what constitutes violence paints a multifaceted portrait:

> … the multifactorial nature of violence itself, with presumably causal factors, most with relatively small contributions to overall risk, having been identified at the individual, interpersonal, community, and societal levels… Individual-level factors include both demographic and biological variables, with violent behavior demonstrating moderate heritability. Violence differs in its goals (instrumental versus impulsive violence), its relationships between perpetrators and victims (intimate versus stranger violence), its nature (physical versus sexual violence), and its relationship to underlying mental disorders, if any. Indeed, even when perpetrated by people with serious mental disorders, only a small proportion of violence appears to be linked directly to symptoms of the disorders themselves. (Appelbaum, 2019, p. 678)

Through the centuries, humans have been faced with the challenge of dealing with complex potentials for violence, often with little advance notice. Creepiness in itself may not be immediately signaling a threat to individuals, but when it is associated in some way with the potential for violence, danger, or unpredictable behavior it becomes a critically-salient aspect of an entity, whether human or technological.

Social Robotics: Bridging Humans, Robots, and AI

The incredible proliferation of advanced information and communication technologies has often lead to some damaging failures as well as future successes (Lee et al., 2021), and the many accidents and fiascos involving robotics and autonomous systems described in the second chapter underscore this point. However, in certain circumstances, interactions with robots and other AI-enhanced technologies serve to construct an engaging and even fulfilling experience in which humans can exercise forms of technological competence. "Social robotics" approaches are special segments of robotic research and development efforts, with emphases on human-robot coordination and communication, which is designed to aid in establishing and sustaining successful and fulfilling human-robot interaction. Mejia and Kajikawa's (2017) bibliometric research on social robotics uncovered eight major "clusters" or themes of research: robots as social partners, human factors and ergonomics on human-robot interaction, robotics for children's development, swarm robotics, emotion detection, assessment of robotic surgery, robots for the elderly, and telepresence and human-robot interaction in rescue robots. With all of these dimensions, this book can only provide some facets of the growing field of social robotics, which includes aesthetic as well as more practical aspects; for example, people are interacting with robots that are designed to exhibit feelings, whether as service robots in the workplace or sex robots in intimate spaces (Yam et al., 2020). Mazzei et al. (2021) describe social robotics as exhibiting exponential growth as a research arena and building on a "multidisciplinary scenario, creating a need for collaboration between different fields" (p. 321). However, these social robotics efforts have been critically analyzed for their potential Machiavellian strategies (Postnikoff, 2020; Sanoubari et al., 2019) as well as potential security risks (Wolfert et al., 2020). For instance, in *My Robot Gets Me: How Social Design Can Make New Products More Human*, Carla Diana (2021) promotes the design of humanoid consumer robots to be responsive in "delightfully subtle ways," with implicit hints rather than transparency as guides for users as to what is involved in interactions. Some implicit design approaches are also being applied to the development of autonomous vehicles, with initiatives to develop design features that fold in seamlessly with other human interaction patterns (Ju, 2015). These implicit interaction routines can include "dark patterns" in which developers intend to obtain some sort of advantage over the user or nudging them to some form of desired behavior.

Along with expressions of feelings by robots and other AI-enhanced entities, the perceived levels and kinds of robot intelligence are often factors in some social robotics approaches and other kinds of human-robot interactions that emerge. Apparent superiority and control of humans by robots can be expressed in various fashions as well as structures that express human inferiority and subservience. Robot competitions have had a substantial impact on the development of robotics, supposedly providing "ideal benchmarks" for robotics research (Behnke, 2006). The contests and comparisons that have supposedly demonstrated to humans that they are inferior to AI-enhanced entities have been going on for decades, from the US quiz show *Jeopardy* battles between IBM's *Watson* and human contestants and chess matches in which computers faced off against human chess masters (Oravec, 2017). For example, in a demonstration of robotic prowess and an implicit comparison with human athletes, Toyota's basketball robot shot series of perfect free throws at the 2020/2021 Tokyo Olympics (Pesce, 2021). Prospects for AI-enhanced systems to replace human journalists in many contexts have also been seriously presented (Miroshnichenko, 2018), with the AI tools reportedly providing more consistent results at a lower cost.

Use of contests in society for establishing and maintaining a social order has gone on for centuries: contests have often been used to demean certain individuals systematically and to support particular unequal social structures (Kohn, 1992). Contests can occur on smaller and more intimate levels, though, as when nurses have to prove that they do better jobs than robots in certain settings (Clinton et al., 2018). The status and superiority established through contests, demonstrations, and other means can play important roles in framing legal and social responsibility, for example, affecting how humans assign accountability to robots versus humans for perceived failures (Lei & Rau, 2021). Since questions as to whether human drivers are less effective than autonomous vehicle technologies are being addressed, governmental agencies as well as the public need to be aware of the tendencies to set up such contests rather than reason on broader and more systemic levels.

Social robotics as a research and development field is entering economic and social climates in which there are many stresses and in which political and social polarizations rather than coordinations are often manifested (Martherus et al., 2019). The potential negative impacts of robotics can be amplified by other pressing societal concerns. For example, Liu and Portes (2020) link the development and proliferation of robots with societal immigration patterns, asking "is the absence of immigrants linked to the rise of automation?" (p. 2723).

The "Brighter Side"?: Constructing Positive Aspects of Robotics

Despite the unsettling concerns discussed in this book, the numbers of robots, self-driving cars, and other AI applications have grown in the past decades, with sizable advances on a variety of fronts, though considerable failures in others (Borjas & Freeman, 2019). Robotic technologies have been implemented in households and fast-food establishments as well as large-scale factory settings; automated vehicles can be found traversing many common transportation routes. Although some individuals have pointed out the deficiencies in robotic companion pets (Sparrow, 2002), their use is becoming widespread in elderly care centers in many nations and they are even being considered for long-term companionship in space travel (Engler et al., 2018). However, the corporate and governmental tendencies to use glitzy marketing and advertising techniques to gloss over difficulties with robotics and AI can produce "hype" instead of the nuanced information that consumers, workers, and household members need to use robots and AI effectively and safely. (The history of AI hype is discussed in an upcoming chapter.) The robotics and AI revolution have largely "been driven by private companies in their own interest" (Slaughter, 2021) along with the sizable creative input of users and implementers; thus, the difficulties of solving some of the social problems involved may often not seem relevant to these organizations and individuals. Some of the concerns individuals have may indeed shift as the "creepiness" they experience is normalized; however, this normalization process may have its own perils for society as robot and AI behaviors are deemed acceptable that can result in some powerful damage.

Advocates of robots and other AI entities might indeed complain about the negative slant of this book, how it emphasizes the unfortunate aspects of human-robot interactions. The book is not intended to paint robotics as uniformly evil or corruptible. There are many useful implementations of these entities that have enhanced individuals' lives, with more to come. For example, by touching the robot PARO, some individuals report reduced pain levels, which is obviously a kind of benefit (Geva et al., 2020); some social robots have served useful roles in educational settings worldwide (Belpaeme & Tanaka, 2021). However, the deaths, injuries, disruptions, and financial losses that are also associated with robotics, automated vehicles, and other AI-enhanced entities should not be dismissed as aberrations, and the related fears and anxieties should not be overlooked in the effort to promote the expansion of a particular technological approach. This negative information can provide clues to whether or how more effective and humane robots and other autonomous entities should or can be developed for everyday workplace and community functioning (Honig et al., 2020). Unfortunately, these unsettling insights can also be used in opportunistic ways in designing entities that are effectively and intentionally frightening. For example, such insights can inform officials and agency heads as to how more effective police and military robots can be designed and how AI-enhanced interrogation devices can be developed, both of which could have significant implications for human rights. The insights can also be used to develop sex robots that are potentially disturbing and triggering to some individuals yet attractive in their depravity to others.

Robotics and AI Designers and Implementers: From Evil Geniuses to Doting Parents to Protesters

The robotics and AI industries comprise a vast assortment of different kinds of organizations, from tiny startups to major international corporations. The last decades have seen major shifts in emphasis, toward support of everyday consumer tasks (Boesl et al., 2019) along with major manufacturing ventures:

> We have used robots for more than six decades to empower people to do things that are typically dirty, dull and/or dangerous. The industry has progressed significantly over the period from basic mechanical assist

systems to fully autonomous cars, environmental monitoring and exploration of outer space. We have seen tremendous adoption of IT technology in our daily lives for a diverse set of support tasks. Through use of robots we are starting to see a new revolution, as we not only will have IT support from tablets, phones, computers but also systems that can physically interact with the world and assist with daily tasks, work, and leisure activities. (Christensen et al., 2021, p. 301)

The relationships of robot persona to their designers and developers can provide additional dimensions of the robots' characterizations. The "Frankenstein" scenario in which a troubled genius creates an entity that goes on to do destructive things has had a centuries-long impact, characterizing a tumultuous and transformative relationship between creator and creation (Shelley, 1818). Two centuries later, "Franken-algorithms," sets of unpredictable computer processes and routines, are providing comparable kinds of concerns (Smith, 2018). The developers of these creations have often been assigned "magical powers" in how they are characterized in marketing efforts, literature, and other forms of public discourse (Kelly, 2018, p. 69).

In considering the dark and creepy sides of robotics, strong and sometimes unflattering linkages to US high-tech culture have been stressed by some commentators: "It should come as no surprise, then, that *the character of the robot has predominantly* been the creation of a small subset of Americans: middle- and upper-class white men" (Abnet, 2020, p. 8). Although the demographic composition of the roster of researchers and engineers who work on robots and AI is changing in character (and becoming more diverse), understanding some of the motivations and approaches of the people who worked on these technologies in decades past may be of value. The story of the late Diana Forsythe in the 1980s is of special value, as her AI work was apparently marginalized (Forsythe, 2001; Oravec, 2004). However, the number of "celebrity" robotics and AI developers (and associated biographical and motivational information) is small; associations of specific individuals with computer platforms and inventions have yielded some luminaires, but relatively few in relation to entertainment, sports, and political fields. For example, Mark Zuckerberg from Meta Corporation and Craig Newmark from Craiglist (Oravec, 2014) have left sizable personal imprints on high-tech arenas, but many developers of major platforms are without substantial name recognition. The other, less well recognized, developers and designers are often best understood through descriptions and discussions of their products and corporate affiliations and their expressed motivations in producing the entities and being part of the organizations involved. For instance, in *Anatomy of a Robot*, engineer Charles M. Bergren (2003) describes the particular joys of building robots, activities he claims are significantly different from those of many other forms of engineering:

> And passion makes it all possible! Personally, I feel it's just as important to understand why I'm doing these things as it is to actually do them. I am old enough to realize that I will never fully understand my motives, nor should I. If I really found out exactly why I liked this field, the fantasy would probably be gone and I'd have to move on to something else. Something is deliciously evil about trying to construct robots to carry out our bidding when we do not even know our own wishes and desires. (p. 2). Building robots is much like going into battle. You can do great damage coming straight out of the gate and swinging swords, but it takes planning to make sure only the enemy gets cut. (p. xii)

In comparably glowing terms, Kang (2021) relates how some engineers who once worked on robotics projects for space missions are attempting with considerable enthusiasm to transform the pizza-making process (including some of its more nuanced human touches) into one that is amenable to robotics, essentially enabling robots to trespass into a formerly-human realm.

With the potentials for robot-related pain and disruption outlined in this book in mind, individuals should be concerned about the possibilities for opportunism, neglect, manipulation, and even sadism on the part of the designers and implementers of the robotics and other AI systems they utilize, even moving beyond the "deliciously evil" motivation that Bergren describes above. In this context, "sadism" involves some amount of pleasure from the infliction of pain upon entities in which one has some amount of association or personal identification (Baumeister & Campbell, 1999). In contrast, some other developers are beginning to protest the purposes to which their creations are being applied. To those individuals, it would be a serious affront to work in organizations in which "killer robots" are being developed or in which autonomous entities are characterized in toy-like or childish ways that may contribute to their users' infantilization. Protests of an assortment of AI-related developments have conveyed activist approaches and themes among some developers and users (Young et al., 2021). The powerful forms of personal control that AI-enhanced systems can institute are of a type reminiscent of the social experiments of past decades (such as those conducted by Stanley Milgram) that resulted in some significant psychological damage to their participants (Nicholson, 2011). Krafft et al. (2021) have proposed that AI-enhanced systems be included in "technology audits" by community advocates and activists. The "responsible robotics" movement among developers is another effort to understand and mitigate potential damages related to robotics (van Wynsberghe, 2021). Some concerned developers and researchers have focused on the "human rights" aspects of the robot-human-AI relationship (Aizenberg & van den Hoven, 2020); analyses of the racial and gender biases of AI-enhanced systems have also been a focus (Fazelpour & Danks, 2021; Mac, 2021).

Robots and other AI-enhanced entities are increasingly mass produced, ubiquitous items that are being viewed in light of how their ready availability (and user-inspired adaptability, in some cases) can affect how everyday tasks are conceptualized. The images and narratives of the individuals who help to create robots and AI innovations are becoming more prevalent and are playing increasing roles in how robotic and other autonomous entities are characterized in popular discourse. Although some of the developers of robots construct them so as to be destructible (as in the "student robot club" efforts in some high schools), many developers profess a strong personal interest in creating entities that have a continuing existence and even a considerable societal impact. For example, the visions of early robotics pioneers provided direction for a number of research efforts. Noted engineer Joseph Engelberger inspired research efforts to create an "Elderly Care Giver," a "multitasking personal robot assistant for everyday care needs in old age" (Van Aerschot & Parviainen, 2020, p. 247); Engelberger's book *Robotics In Practice* (1983) inspired considerable R&D activity. In research on robotics, Brandao (2021) attempted to identify the perspectives and values reflected in technical robotics papers, identifying "multiple visions used by robotics researchers in their published work: robots in our daily (personal) lives, robots cooperating with humans, replacing humans, doing dangerous jobs, assisting the elderly, becoming human-like, and respecting culture" (p. 6). Brandao's work is useful here but may overlook the corporate efforts that are designed to make interacting with technologies addictive (Eyal, 2014) or that encourage people to cooperate in planned obsolescence strategies in which new, supposedly-updated technologies are required and older technologies are destroyed.

The "evil geniuses" or "mad scientists" described by Grazier and Cass (2015), Skal (1998), and Webb (2017) are limited in number. Celebrity scientists and developers are limited as well; for every Elon Musk of Tesla Corporation in the US, there are countless others who are part of the research and development infrastructure of robotics and AI. Having celebrity does not automatically entail being able to procure the funding needed for research and development; for example, Douglas Engelbart, inventor of the mouse, was denied substantial financing for decades of his life (Foremski, 2013). Those involved in this R&D community have often avoided ethical discourse related to their efforts, with some notable exceptions described in the last chapter:

> … robot creators tend to be engineers, programmers, and designers with little training in ethics, human rights, privacy, or security. In the United States, hardly any of the academic engineering programs that grant degrees in robotics require the in-depth study of such fields. (Nourbakhsh, 2015, p. 23)

The major and most important audiences for new robotic creations are often other robot and AI developers and researchers, some of whom can help pave the way to larger grants or more lucrative commercial contracts. The "demo or die" frontstage presentations of robotic developments have served an important role for the past decades in showcasing technological advances:

> Just as other scientists and engineers, roboticists routinely present their work to academic peers and sponsors, as well as to potential customers and the lay public – and they are expected to do so in an increasingly professionalized manner. They stage live and video demonstrations, are involved in science communication efforts, and those who are (also) entrepreneurs have to engage in public relations and marketing as well. "Researchers always have something to sell. … Those working in academia are looking for talk invitations." (Voss, 2021, p. 75)

As various social and public policy concerns about robots emerge, communications among relevant parties as to how to proceed in regulation and oversight of robotics and other AI developments are often stalled. Fosch-Villaronga and Heldeweg (2018) argue that "there is no formal communication process between robot developers and regulators from which policies could learn" (p. 113). New kinds of professions are emerging as robotics and AI entities require various forms of prepping and technical support (Share & Pender, 2018), which adds new parties to the mix and creates more potentials for positive social change as well as for negative forms of disruption.

Robot Failures, Orphans, and "Robotic Winters"

Robotic and autonomous vehicle product failures are accumulating, playing special roles in reasoning about the technologies' applications in society. The "orphaning" and abandonment of specific items have often brought with it larger questions about technological directions and management (Gandal et al., 1999). Technologies can be orphaned at a variety of stages: an orphan can be "one where the technology is not commercialized, after its development. Yet, another view sees a technology orphan as one, which after introduction is not further used, for any number of reasons" (Barbonis, 2006). The pace of research and development growth in robotics and AI have often been dissatisfying for many corporate leaders and financial supporters, leading to "robot winters" and "AI winters" in which research and development funding can be scarce (Floridi, 2020; Oravec & Travis, 1992). The term "AI winter" was reportedly coined in 1984 by AI researchers Marvin Minsky and Roger Schank to characterize impending deaths of resources invested in AI as well as their implications for certain emerging research agendas and the future directions of graduate students and junior researchers. Although robots and AI-enhanced entities such as autonomous vehicles have a growing societal presence, the rate of their proliferations has stalled in relation to the projections of their advocates in a number of venues:

> Yet there is still no implementation on a large scale. There are several reasons to explain the slow adoption. Robots, and particularly social robots, are generally considered expensive for their actual capabilities, and their cost–benefit tends to be disadvantageously compared to that of voice assistants. Additionally, there are sociocultural factors in Western societies that explain this reluctance when compared to Japan, for example. One might be robot aversion, influenced by science fiction, and arguably rooted in Judeo-Christian beliefs that associate the creation of 'human-like' creatures to an act of hubris. (Aymerich-Franch, 2020, p. 364)

Some recent examples of failures in robotics development and implementation include the US corporation Walmart's jettisoning its plans to use robots to scan shelves as part of supply chain management and inventory control (Nassauer, 2020). The response of human co-workers to the efforts of robots has often been imbued with fear and anxiety, so such a move by Walmart may be welcomed by many of its workers. The following account exemplifies these feelings:

> Marty actually moves around the store unassisted, scanning the floors for spills and trip hazards, which are reported to human workers by making a beeping noise and communicating to employees after paging them. He also has the ability to scan shelves for missing items, do price checks and much more. Here's the bad news: Marty looks weird and is unnerving. He's google eyed, seems to follow you around, and is very —well—robotic. Time will tell whether or not Marty is a success. Maybe because it's one of our first experiences with robotic technology and we're just not used to it, Marty will probably scare the crap out of you the first time you see him. (Franco, 2021, para. 2)

SoftBank Corporation's "Pepper" humanoid robot was introduced in 2014 and marketed in 2015 (Inada, 2021); it was characterized as being able to read and respond to human emotions. Robotics pioneer Noel Sharkey is quoted as saying that "Pepper did a lot to harm genuine robotics research by giving an often false impression of a bright, cognitive being that could hold conversations… it was mostly remote-controlled with a human conversing through its speakers. Deceiving the public in this way is dangerous and gives the wrong impression of the capabilities of AI in the real world" (Pymnts, 2021). The "social robot for the home" Jibo met a similar fate to Pepper's:

> Jibo launched as "the first social robot for the home" and was designed to interact and communicate with those around it. The internet-connected device used voice- and facial-recognition smarts to forge relationships with familiar folks, and could read out messages from family and friends, while a built-in camera set it up as the family photographer. A display for a face allowed for a more interactive video-call experience, too, with Jibo's swiveling head able to move around to involve everyone in the room.
>
> The robot started life as a crowdfunding campaign, securing more than $3 million in 2012. Venture capital funding saw an additional $70 million

pumped into the project, but despite its best efforts, the company still couldn't make a success of it… a slew of factors, including its high price, delayed shipments, poor reviews, and the arrival of Amazon's considerably cheaper Echo smart speaker and other similar devices. (Mogg, 2021)

Many more consumer robotic ventures may turn out to be failures in the marketplace, perhaps providing clues as to how and whether robotic "creepiness" has gone too far and has resulted in audiences that will not buy the product. Consumers who bought and started using the technologies involved can often be stuck with "orphans" that are not fully supported and updated by their developers, providing further (and very dangerous) problems involving security. Barbonis (2006) relates that "It might also be interesting for firms to undertake an audit of the actual cost of developing a technology, and maintaining it until such time it is orphaned, as also the intangible cost and other social consequences of orphaning" (p. 472). Robotic failures provide special concerns because of the amount of investment consumers can make in altering their working environments and essential processes in order to accommodate the robots, only to have entities in their midst that can attract viruses and other security breaches.

As fears for some kinds of "robot winters" emerge, unfortunate decisions can be made about how to allocate research and development recourse in the area, favoring strategic agendas that are immediately lucrative instead of promising in terms of the overall direction of the industry. Autonomous cars are also producing problematic issues concerning their capabilities and uses as they are involved in lethal incidents (described in the next chapter); basic parameters of the vehicles' safety are in question (Stilgoe, 2018). Some of the vehicles are generating unsettling popular narratives that are similar to those about robots, such as "Self-driving cars are a no-go, so let's all pretend flying cars are coming" (Lykiardopoulou, 2021). The self-driving car project Waymo's autonomous vehicles have reportedly encountered manipulations and harassment by humans that have led to many accidents (Reagan, 2021). A broad assortment of optimistic, positive accounts of autonomous vehicles is being generated and disseminated by car companies, however, which may eventually serve to overwhelm negative messages and unsettling news accounts.

Some Conclusions and Reflections

Attitudes and approaches toward robotics, autonomous vehicles, and AI can vary significantly and can even be seen as "polarizing" in some ways as these entities' dark and light sides are contrasted. Robots can often serve as a *Rorschach* test (inkblots that are used as projective psychological tests) to elicit individuals' perspectives on certain societal and psychological dimensions. Many of the various "dark side" issues and concerns described in this book are significant in terms of personal wellbeing as well as overall societal and organizational health. However, working with and being monitored by robots and AI are becoming compulsory parts of our lives, so advising people to walk away from robots if they are disturbed or upset by them will not be sufficient in many contexts. As discussed in the following chapters, these dark side issues can be downplayed in many social contexts in favor of an approach that makes robotics and AI inevitable and their applications somehow appropriate and even necessary for various settings, bolstered by economic arguments and various "contests" in which human efforts are compared (often unfavorably) with those of robots. The unfortunate outcomes of the proliferation of robots and other AI-enhanced entities (including fear and addiction) can be used in opportunistic ways by various developers, implementers, and users. For example, the generalized fear that people may acquire of police and military robots (described in an upcoming chapter) may indeed not be proportional to the risks involved in a particular, limited circumstance; however, this fear can be a factor in social control in various communities and regions, changing people's routine habits and public behaviors to be in synch with their assumptions about possibly-punitive robot activity. Addiction to sex robots and even to some domestic robots can also be used opportunistically by those who market them. For example, although such behaviors as "hoarding robots" have not yet emerged on a large scale, addictive patterns may soon extend to such accumulative tendencies. Efforts to foster addiction to technologies by corporations can serve to expand sales and increase the amount of intimate information collected about users.

The chapters to come introduce questions that probe the scope of human agency within robotic and AI realms, such as how far will people go in allowing themselves to be socialized or controlled by robots and other AI-enhanced entities? As related in the upcoming chapter on military and police robots, some interactions with autonomous entities will be made unavoidable and capable performance in these encounters will be necessary for personal survival. If an individual is ordered to "halt" by a police robot, various ranges of outcomes can be projected by the individual and the robot with a certain set of uncertainties. If the police robot malfunctions either through the indeterminacy of its programming or by the malicious intent of another human, the instance involved can morph into a tragic situation unless a human's awareness and common sense can somehow intercede. Whether deception, abuse, manipulation, and other dark side issues will become acknowledged challenges for designers and users to overcome, or become normalized as underhanded and surreptitious tools for opportunistic utilization, is yet to be determined. Possibly more difficult is analyzing the impacts that robots and AI are having on human character. In order to play within the rules and constraints of the robot designer's game, individuals generally must accept some basic assumptions and allow oneself to be constrained by various limitations; many insightful users will be able to extend these games somewhat, however, including some creative twists. Analyzing human interaction with robots will provide clues to how and whether human responses can be modified, which is why the experimental results of the "social robotics" movement will be useful whatever one's attitudes toward robotics and AI.

What will the workplaces and communities of the future be like in terms of their implementations of robotics, autonomous vehicles, and AI? Organizational and community settings have always had "others" that are unfortunately designated, whether they are migrants or people of certain demographic categories. Some of the aggression that might otherwise have been directed toward women, individuals with disabilities, or members of certain groups may soon be directed toward robots (as described in an upcoming chapter), as the robots become symbols of job loss and human diminution. Will we walk past the "Don't Bully the Robots" sign, which is strategically placed so as to forestall signs of such anti-robot aggression? The American Society for the Prevention of Cruelty to Robots is a pioneer in framing these issues. It was created by Peter Remine in 1999 in order to underscore concerns involving dysfunctional human and robotic interaction (Belk, 2018). The presence of robots (and related anti-robot aggressions) could indeed deflect tensions that might otherwise be directed toward real-life individuals, and may thus relieve some forms of community, workplace, and even household disquiet, though with problematic results at other levels. However, some of the kindness and acceptance that should be granted unquestionably to humans may also be inappropriately directed toward robots. Efforts to increase the empathy of humans toward robots may serve to misdirect the often-finite resources of human patience and sympathy. Whether we treat robots and AI-related entities as having some special powers or presence based on assumptions about their magical powers or just because they are cute, we need to make and reinforce a strong and binding distinction between humans and robots, not work to muddy the discrepancies.

People who need to make decisions about robot, autonomous vehicle, and AI implementations can face a complex and confusing set of technical terms as well as science fiction and marketing images. Efforts to develop a "critical literacy" of emerging robot and AI initiatives can encounter difficulties as the growing assortment of AI-enhanced entities is considered. For example, the very notion of "autonomous" operations in the context of robotics and automated vehicles is complex and changing with advances in AI. The anthropomorphic dimensions of many robots (such as a humanlike face or specific gender or racial assignments) are sometimes explicit and intrinsic aspects of design and in other contexts are added features provided for advertisement or entertainment. Various anthropomorphic dimensions are also being assigned to autonomous vehicles, such as voice quality and expressions on a car's grill; design theorist Donald Norman (1992) proclaimed decades ago that "turn signals are the facial expressions of automobiles." Despite the early stages of knowledge about the impacts of robots and AI-enhanced entities in social arenas, social engineering of human-robot and human-autonomous vehicle interactions and the environments in which they are conducted are becoming considerable aspects of managerial and public policy efforts. Individuals often engage in comparable interactions in their households as robots and AI-enhanced entities take roles in everyday activities. As discussed in chapters to come, the intimate and ethical aspects of our lives are also being influenced by robots and AI. The "sexual engineering" of how individuals characterize and use sex robots is proceeding in many computer industry settings (described in an upcoming chapter), as is the production of fear-eliciting military robots and addiction-inducing commercial entities. The development of "ethical robots" can indeed be attempted (entities programmed to abide within the constraints and affordances of particular ethical rules), but the potential for them to be reprogrammed by others or to malfunction in their laudatory purposes can erode their intended usability (Vanderelst & Winfield, 2018). People who are afforded more transparency about robot and AI engineering and implementation, as well as who become aware of their own superstitions and fears about robots and AI, may be more balanced and evenhanded in their everyday decision making concerning these entities.

REFERENCES

Aizenberg, E., & van den Hoven, J. (2020). Designing for human rights in AI. *Big Data & Society*, *7*(2). https://doi.org/10.1177/2053951720949566

Appelbaum, P. S. (2019). In search of a new paradigm for research on violence and schizophrenia. *American Journal of Psychiatry*, *176*(9), 677–679. https://doi.org/10.1176/appi.ajp.2019.19070678

Autor, D. (2015). Why are there still so many jobs? The history and future of workplace automation. *Journal of Economic Perspectives*, *29*(3), 3–30. https://doi.org/10.1257/jep.29.3.3

Baccarella, C. V., Wagner, T. F., Kietzmann, J. H., & McCarthy, I. P. (2018). Social media? It's serious! Understanding the dark side of social media. *European Management Journal*, *36*(4), 431–438.

Baeva, L. V. (2018). Cyborgization: Pros and cons. In *Androids, cyborgs, and robots in contemporary culture and society* (pp. 138–150). IGI Global.

Barbonis, P. A. (2006, September). Towards some theory of technology orphaning: A retrospective exploratory study. In *2006 IEEE International Engineering Management Conference* (pp. 468–472). IEEE Press.

Barlas, Z. (2019). When robots tell you what to do: Sense of agency in human- and robot-guided actions. *Consciousness and Cognition*, *75*. https://doi.org/10.1016/j.concog.2019.102819

Barrett, M., Oborn, E., Orlikowski, W. J., & Yates, J. (2012). Reconfiguring boundary relations: Robotic innovations in pharmacy work. *Organization Science*, *23*(5), 1448–1466.

Baumeister, R. F., & Campbell, W. K. (1999). The intrinsic appeal of evil: Sadism, sensational thrills, and threatened egotism. *Personality and Social Psychology Review*, *3*(3), 210–221.

BBC News. (2021). Thousands of orders cancelled after Ocado robot fire. *BBCNews.com*. https://www.bbc.com/news/business-57883332

Beck, B. (2020). Infantilisation through technology. In *Technology, anthropology, and dimensions of responsibility* (pp. 33–44). JB Metzler.

Behnke, S. (2006, October). Robot competitions-ideal benchmarks for robotics research. In *Proceedings of IROS-2006 Workshop on Benchmarks in Robotics Research*. Institute of Electrical and Electronics Engineers (IEEE) Press.

Belk, R. (2016). Understanding the robot: Comments on Goudey and Bonnin (2016). *Recherche et Applications En Marketing (English Edition)* (Sage

Publications Inc.), 31(4), 83–90. https://doi.org/10.1177/2051570716658467

Belk, R. (2017). Magical machines meet magical people. In E. P. Becerra, R. Chitturi, M. Cecilia Henriquez Daza, & J. Carlos Londoño Roldan (Eds.), *ACR Latin American Advances* (Vol. 4, pp. 28–30).Association for Consumer Research. https://www.acrwebsite.org/volumes/1700019/la/v4_pdf/LA-04

Belk, R. (2018). Ownership: The extended self and the extended object. In *Psychological ownership and consumer behavior* (pp. 53–67). Springer.

Bellomi, J. J. (2019). *Darker and darker still: Media-technology, darkness narratives, and fear.* University of California, Santa Barbara. https://www.proquest.com/openview/bbf413e904a44637bf9360392b2e0a65/1?pq-origsite=gscholar&cbl=18750&diss=y

Belpaeme, T., & Tanaka, F. (2021). Social robots as educators. In *OECD digital education outlook 2021: Pushing the frontiers with artificial intelligence, blockchain and robots* (pp. 143–157).

Bergren, C. M. (2003). *Anatomy of a robot.* McGraw-Hill.

Beware: Robots do harm humans! (2017). *Keller's OSHA Safety Training Newsletter, 24*(12), 4.

Bindley, K., & Elliott, R. (2021, May 20). Tesla drivers test Autopilot's limits, attracting audiences—and safety concerns. *Wall Street Journal.* https://www.wsj.com/articles/tesla-drivers-test-autopilots-limits-attracting-audiencesand-safety-concerns-11621503008

Blumenberg, H. (1993). Light as a metaphor for truth: At the preliminary stage of philosophical concept formation. In D. M. Levin (Ed.), *Modernity and the hegemony of vision* (pp. 30–62). University of California Press.

Boesl, D. B., Bode, M., & Greisel, S. (2019, April). Successful consumer robotics beyond science fiction-use case based requirements engineering for product development of a consumer robot. In *2019 IEEE 23rd International Conference on Intelligent Engineering Systems (INES)* (pp. 000159–000164). IEEE. https://ieeexplore.ieee.org/abstract/document/9109533

Borjas, G. J., & Freeman, R. B. (2019). From immigrants to robots: The changing locus of substitutes for workers. *RSF: The Russell Sage Foundation Journal of the Social Sciences, 5*(5), 22–42.

Bradbury, M. (2000). *To the hermitage.* Pan Macmillan.

Brakman, S., Garretsen, H., & van Witteloostuijn, A. (2021). Robots do not get the coronavirus: The COVID-19 pandemic and the international division of labor. *Journal of International Business Studies, 52,* 1215–1224. https://link.springer.com/article/10.1057/s41267-021-00410-9

Brandao, M. (2021). *Normative roboticists: The visions and values of technical robotics papers.* https://www.martimbrandao.com/papers/Brandao2021-roman-visions.pdf

Braun, B. (2000). The X-Files and Buffy the Vampire Slayer: The ambiguity of evil in supernatural representations. *Journal of Popular Film and Television, 28*(2), 88–94.

Brink, K. A., Gray, K., & Wellman, H. M. (2019). Creepiness creeps in: Uncanny valley feelings are acquired in childhood. *Child Development, 90*(4), 1202–1214.

Brown, D. (2021, June 18). An autonomous ship's first effort to cross the Atlantic shows the difficulty of the experiment. *Washington Post*. https://www.washingtonpost.com/technology/2021/06/18/mayflower-ibm-autonomous-ship/

Buck, A. (2020. September 9). Is AI a god of the future or the present? *European Academy of Religion and Society*. https://europeanacademyofreligionandsociety.com/news/is-ai-a-god-of-the-future-or-the-present/

Calo, R. (2016). Robots as legal metaphors. *Harvard Journal of Law & Technology, 30*(1), 209–239.

Carney, D. (2021, November 29). The 6 terms you need to know to understand self-driving cars. *Popular Science*, https://www.popsci.com/technology/autonomous-vehicles-explained/

Casey, B., & Lemley, M. A. (2019). You might be a robot. *Cornell Law Review, 105*, 287–361.

Casey, M. (2015, May 14). Maybe artificial intelligence won't destroy us after all. *CBS News*. https://www.cbsnews.com/news/maybe-artificial-intelligence-wont-destroy-us/

Cave, S., & Dihal, K. (2018). Ancient dreams of intelligent machines: 3,000 years of robots. *Nature, 559*(7715), 473–475.

Choo, K. (2014). Hyperbolic nationalism: South Korea's shadow animation industry. *Mechademia, 9*, 144–162.

Christensen, H., Amato, N., Yanco, H., Mataric, M., Choset, H., Drobnis, A., ... & Sukhatme, G. (2021). A roadmap for US Robotics—From Internet to Robotics 2020 Edition. *Foundations and Trends® in Robotics, 8*(4), 307–424.

Chronaki, A., & Alimisi, R. (2010). Engaging young children to "control" technology: Emotion, negotiation, agency. In *Workshop Proceedings of International Conference on Simulation, Modeling and Programming for Autonomous Robots (SIMPAR 2010) Darmstadt, Germany, November* (pp. 15–18).

Clarke, A. C. (1962). Hazards of prophecy: The failure of imagination. *Profiles of the Future, 6*(36), 1.

Clinton, M., Madi, M., Doumit, M., Ezzeddine, S., & Rizk, U. (2018). "My greatest fear is becoming a robot": The paradox of transitioning to nursing practice in Lebanon. *SAGE Open, 8*(2), 2158244018782565.

Coates, J. F. (2000). Looking ahead: There is no Joy in my life. *Research Technology Management, 43*(6), 6–8.

Coeckelbergh, M. (2020). *AI ethics*. MIT Press.

Collins, K. (2021, January 4). At CES, Hyundai sees Boston Dynamics robots as a part of the metaverse. *C/net*. https://www.cnet.com/tech/computing/at-ces-hyundai-sees-boston-dynamics-robots-as-a-part-of-the-metaverse/

Condliffe, J. (2019). The week in tech: Some workers hate robots. Retraining may change that. *New York Times*. https://www.nytimes.com/2019/07/19/technology/amazon-automation-labor.html

Crane, L. (2018, August 15). *Replacing your boss with a cruel robot could make you concentrate more*. https://www.newscientist.com/article/2176798-replacing-your-boss-with-a-cruel-robot-could-make-you-concentrate-more/#ixzz6uS27ZXq8

Cumbley, R., & Church, P. (2013). Is "big data" creepy?. *Computer Law & Security Review, 29*(5), 601–609.

Damiano, L. (2021). Homes as human–robot ecologies: An epistemological inquiry on the "domestication" of robots. *The Home in the Digital Age* (pp. 80–102). Routledge.

D'Cruz, P., & Noronha, E. (2021). Workplace bullying in the context of robotization: Contemplating the future of the field. In *Concepts, approaches and methods* (pp. 293–321). Springer. https://doi.org/10.1007/978-981-13-0134-6_22

Del Casino Jr, V. J., House-Peters, L., Crampton, J. W., & Gerhardt, H. (2020). The social life of robots: The politics of algorithms, governance, and sovereignty. *Antipode, 52*(3), 605–618.

Devlin, K. (2018). *Turned on: Science, sex and robots*. Bloomsbury Publishing.

Diana, C. (2021). *My robot gets me: How social design can make new products more human*. Harvard Business School Press.

Duan, H., Li, J., Fan, S., Lin, Z., Wu, X., & Cai, W. (2021, October). Metaverse for social good: A university campus prototype. In *Proceedings of the 29th ACM International Conference on Multimedia* (pp. 153–161). ACM Press.

Effoduh, J. O. (2021). The legitimization of customized sex robots in the age of COVID-19. *Intellectual Property Journal, 33*(2), 161–181.

Emerson, R. W. (1870). *Society and solitude: Twelve chapters. By Ralph Waldo Emerson (Low's Copyright Cheap Editions of American Books.)*. Sampson Low.

Engelberger, J. (1983). *Robotics in practice: Management and applications of industrial robots*. Springer.

Engler, S., Hunter, J., Binsted, K., & Leung, H. (2018, June). Robotic companions for long Term isolation space missions. In *2018 15th International Conference on Ubiquitous Robots (UR)* (pp. 424–430). IEEE Press.

Eyal, N. (2014). *Hooked: How to build habit-forming products*. Penguin.

Fazelpour, S., & Danks, D. (2021). Algorithmic bias: Senses, sources, solutions. *Philosophy Compass, 16*(8). https://doi.org/10.1111/phc3.12760

Foremski, T. (2013, July 3). The shocking truth about Silicon Valley genius Doug Engelbart. *ZDNet.* https://www.zdnet.com/article/the-shocking-truth-about-silicon-valley-genius-doug-engelbart/

Forsythe, D. (2001). *Studying those who study us: An anthropologist in the world of artificial intelligence.* Stanford University Press.

Fortunati, L., Esposito, A., Sarrica, M., & Ferrin, G. (2015). Children's knowledge and imaginary about robots. *International Journal of Social Robotics, 7*(5), 685–695.

Fosch-Villaronga, E., & Heldeweg, M. A. (2018, October). "Meet me halfway," said the robot to the regulation. In *International Conference on Inclusive Robotics for a better Society* (pp. 113–119). Springer.

Fotheringham, D., Lisjak, M., & Kristofferson, K. (2020). Rage against the machine: When consumers sabotage robots in the marketplace. *ACR North American Advances* (p. 714). https://www.acrwebsite.org/volumes/2662517/volumes/v48/NA-48

Franco, J. (2021, May 21). Creepy robots have arrived at your NJ supermarkets. *NJ1015.com.* https://nj1015.com/creepy-robots-have-arrived-at-your-nj-supermarkets-opinion/?utm_source=tsmclip&utm_medium=referral

Gain, V. (2021, November 2). Meta makes robot 'skin' and sensors to give metaverse a new touch. *Silicon Republic.* https://www.siliconrepublic.com/machines/facebook-meta-robot-skin-sensor-metaverse

Gandal, N., Greenstein, S., & Salant, D. (1999). Adoptions and orphans in the early microcomputer market. *The Journal of Industrial Economics, 47*(1), 87–105.

Gebru, T., & Mitchell, M. (2022, June 17). We warned Google that people might believe AI was sentient. Now it's happening. *Washington Post.* https://www.washingtonpost.com/opinions/2022/06/17/google-ai-ethics-sentient-lemoine-warning/

Geraci, R. M. (2006). Spiritual robots: Religion and our scientific view of the natural world. *Theology and Science, 4*(3), 229–246.

Geva, N., Uzefovsky, F., & Levy-Tzedek, S. (2020). Touching the social robot PARO reduces pain perception and salivary oxytocin levels. *Scientific Reports, 10*(1), 1–15.

Goldenberg, J., Mazursky, D., & Solomon, S. (1999). Creative sparks. *Science, 285*(5433), 1495–1496.

Golds, A. (2021, June 13). Meet the robots inside Amazon's fulfillment center. *CBS News.* https://www.cbsnews.com/news/meet-the-robots-inside-amazons-fulfillment-centers/

GPT-3. (2020, September 8). A robot wrote this article. *The Guardian.* https://www.theguardian.com/commentisfree/2020/sep/08/robot-wrote-this-article-gpt-3

Grazier, K. R., & Cass, S. (2015). Hollywood scientists: Reel and imaginary. In *Hollyweird science* (pp. 47–89). Springer.

Halfacree, G. (2021, June 18). Poltergeist attack could leave autonomous vehicles blind to obstacles—Or haunt them with new ones: First 'AMpLe' concept proves worryingly simple to implement with success. *The Register*. https://www.theregister.com/2021/06/18/poltergeist_autonomous_vehicles/

Haraway, D. (1987). A manifesto for cyborgs: Science, technology, and socialist feminism in the 1980s. *Australian Feminist Studies, 2*(4), 1–42. https://doi.org/10.1080/08164649.1987.9961538

Harris, M. (2021, August 16). Here's what the inevitable friendly neighborhood robot invasion looks like. *Techcrunch*. https://techcrunch.com/2021/08/16/nuro-ec1-operations/

Heffernan, T. (2018). AI artificial intelligence: Science, fiction and fairy tales. *English Studies in Africa, 61*(1), 10–15.

Heffernan, T. (2020). The dangers of mystifying artificial intelligence and robotics. *Toronto Journal of Theology, 36*(1), 93–95.

Hermann, I. (2020). Beware of fictional AI narratives. *Nature Machine Intelligence, 2*(11), 654–654.

Hinks, T. (2021). Fear of robots and life satisfaction. *International Journal of Social Robotics, 13*(2), 327–340.

Honig, S., Bartal, A., & Oron-Gilad, T. (2020, March). Using customers' online reviews to identify and classify human robot interaction failures in domestic robots. In *Companion of the 2020 ACM/IEEE International Conference on Human-Robot Interaction* (pp. 251–253). IEEE Press.

Honig, S., & Oron-Gilad, T. (2018). Understanding and resolving failures in human-robot interaction: Literature review and model development. *Frontiers in Psychology, 9*, 861–882.

Hu, Y. (2021). An improvement or a gimmick? The importance of user perceived values, previous experience, and industry context in human–robot service interaction. *Journal of Destination Marketing & Management, 21*, 100645.

Hughes, J. (1997). *The intelligent other in science, fantasy and horror fiction, 1895 to the present*. http://ieet.org/archive/19990820-BioOther.pdf

Inada, M. (2021, July 13). SoftBank introduced Pepper, its humanoid robot, to the world in 2014 and started selling it the next year. *Wall Street Journal*. https://www.wsj.com/articles/humanoid-robot-softbank-jobs-pepper-olympics-11626187461

Isidore, C. (2021, December 23). Tesla is under federal investigation for letting drivers play video games. *CNN.com*. https://www.cnn.com/2021/12/23/tech/tesla-safety-probe-drivers-play-video-games/index.html

Johnson, B. D. (2014). Violence, death and robots: Going to extremes with science fiction prototypes. *Personal and Ubiquitous Computing, 18*(4), 809–810. https://doi.org/10.1007/s00779-013-0680-0

Johnson, D. G., & Verdicchio, M. (2017). AI anxiety. *Journal of the Association for Information Science and Technology, 68*(9), 2267–2270.

Ju, W. (2015). The design of implicit interactions. *Synthesis Lectures on Human-Centered Informatics, 8*(2), 1–93.

Kang, M. (2011). *Sublime dreams of living machines*. Harvard University Press.

Kang, M. (2021, December 15). Can pizza robots be engineered to have a human touch? *LA Eater*. https://la.eater.com/2021/12/15/22837994/stellar-pizza-robot-machine-hawthorne-los-angeles-automation-technology

Karr-Wisniewski, P., & Lu, Y. (2010). When more is too much: Operationalizing technology overload and exploring its impact on knowledge worker productivity. *Computers in Human Behavior, 26*(5), 1061–1072.

Kassens-Noor, E., Wilson, M., Cai, M., Durst, N., & Decaminada, T. (2021). Autonomous vs. self-driving vehicles: The power of language to shape public perceptions. *Journal of Urban Technology, 28*(3–4), 5–24.

Kelly, N. M. (2018). "Works like magic": Metaphor, meaning, and the GUI in Snow Crash. *Science Fiction Studies, 45*(1), 69–90.

Kim, M. S., & Kim, E. J. (2013). Humanoid robots as "The cultural other": Are we able to love our creations? *AI and Society, 28*(3), 309–318. https://doi.org/10.1007/s00146-012-0397-z

Kirschgens, L. A., Ugarte, I. Z., Uriarte, E. G., Rosas, A. M., & Vilches, V. M. (2018). Robot hazards: From safety to security. *arXiv preprint* arXiv:1806.06681

Klotz, F. (2016). Are you ready for robot colleagues? *MIT Sloan Management Review, 58*(1). https://www.proquest.com/scholarly-journals/are-you-ready-robot-colleagues/docview/1832180555/se-2?accountid=14791

Kohn, A. (1992). *No contest: The case against competition*. Houghton Mifflin Harcourt.

Kumar, P. M., Pandey, H. M., & Srivastava, G. (2021). Special issue on workplace violence prevention using security robots. *Work (reading, MA), 68*(3), 821–823. https://doi.org/10.3233/WOR-203415

Kumar, G., Singh, G., Bhatanagar, V., & Jyoti, K. (2019). Scary dark side of artificial intelligence: A perilous contrivance to mankind. *Humanities & Social Sciences Reviews, 7*(5), 1097–1103.

Kurzweil, R. (2005). *The singularity is near*. Penguin Books.

Laakasuo, M., Palomäki, J., & Köbis, N. (2021). Moral uncanny valley: A robot's appearance moderates how its decisions are judged. *International Journal of Social Robotics*, 1–10.

Lacey, C., & Caudwell, C. (2019, March). Cuteness as a 'dark pattern' in home robots. In *2019 14th ACM/IEEE International Conference on Human-Robot Interaction (HRI)* (pp. 374–381). IEEE Press.

Latour, B. (2005). *Reassembling the social: An introduction to actor-network-theory*. Oxford University Press.

LeBlanc, W. (2021, January 15). Sophia the Robot creators announce plan to mass-produce robots this year. *Ign.com*. https://www.ign.com/articles/sophia-the-robot-creators-announce-plan-to-mass-produce-robots-this-year

Lee, J. L. (2021, August 26). Factbox- Investors are betting on rent-a-robot startups in Silicon Valley. *Yahoo.com*. https://finance.yahoo.com/news/factbox-investors-betting-rent-robot-102730625.html

Lee, J. S., Keil, M., & Wong, K. F. E. (2021). When a growth mindset can backfire and cause escalation of commitment to a troubled information technology project. *Information Systems Journal, 31*(1), 7–32.

Lei, X., & Rau, P. L. P. (2021). Effect of relative status on responsibility attributions in human–robot collaboration: Mediating role of sense of responsibility and moderating role of power distance orientation. *Computers in Human Behavior, 122*. https://doi.org/10.1016/j.chb.2021.106820

Lindley, J., Coulton, P., & Alter, H. (2019). Networking with ghosts in the machine. Speaking to the Internet of Things. *The Design Journal, 22*(Suppl. 1), 1187–1199.

Liu, L., & Portes, A. (2020). Immigration and robots: Is the absence of immigrants linked to the rise of automation? *Ethnic and Racial Studies, 44*(15), 2723–2751. https://doi.org/10.1080/01419870.2020.1849757

McLuhan, M. (1964). *Understanding media: The extensions of man*. MIT Press.

Mac, R. (2021, September 4). Facebook apologizes after AI puts 'primates' label on video of black men. *New York Times*. https://www.nytimes.com/2021/09/03/technology/facebook-ai-race-primates.html

Maldonato, N. M., Valerio, P., Bottone, M., Sperandeo, R., Scandurra, C., Punzo, C., ... & Eposito, A. (2021). The desiring algorithm. The sex appeal of the inorganic. In *Progresses in Artificial Intelligence and Neural Systems* (pp. 607–613). Springer.

Marrone, S. P. (2014). *A history of science, magic and belief: From medieval to early modern Europe*. Macmillan International Higher Education.

Marsh, J. (2019). The uncanny valley revisited: Play with the internet of toys. In *The Internet of toys* (pp. 47–65). Palgrave Macmillan.

Martherus, J. L., Martinez, A. G., Piff, P. K., & Theodoridis, A. G. (2019). Party animals? Extreme partisan polarization and dehumanization. *Political Behavior, 43*, 517–540.

Mash, S. (2021, July 8). Cybersecurity for industrial robots. *Electronics 360*, https://electronics360.globalspec.com/article/16904/cybersecurity-for-industrial-robots

Mathur, A., Acar, G., Friedman, M. J., Lucherini, E., Mayer, J., Chetty, M., & Narayanan, A. (2019). Dark patterns at scale: Findings from a crawl of 11K shopping websites. *Proceedings of the ACM on Human-Computer Interaction, 3*(CSCW), 1–32.

Matyszczyk, C. (2020, April 3). Robots must teach humans about love, says world's most famous robot: For the world's future, robots must show more hope and compassion than humans. So says Sophia the celebrity robot. *ZDnet*. https://www.zdnet.com/article/robots-must-teach-humans-about-love-says-worlds-most-famous-robot/

Matyszczyk, C. (2021, July 17). Amazon has a twisted, clever new idea that you really may not love: Tech companies want to make the world a better place. But look what Amazon wants to do to your neighborhood now. *ZDnet*. https://www.zdnet.com/article/amazon-has-a-twisted-clever-new-idea-that-you-really-may-not-love/

Mayor, A. (2018). *Gods and robots*. Princeton University Press.

Mazzei, D., Chiarello, F., & Fantoni, G. (2021). Analyzing social robotics research with natural language processing techniques. *Cognitive Computation, 13*(2), 30.

McBride, J. (2019). Robotic bodies and the kairos of humanoid theologies. *Sophia, 58*(4), 663–676.

McKay, T. (2017). No, Facebook did not panic and shut down an AI program that was getting dangerously smart. *Gizmodo*. Available at: https://gizmodo.com/no-facebook-did-not-panic-and-shut-down-an-ai-program-1797414922. https://futurism.com/the-byte/roundup-ranks-creepy-robots

Meinecke, L., & Voss, L. (2018). I Robot, You unemployed: Robotics in science fiction and media discourse. *Schafft Wissen. Gemeinsames und geteiltes Wissen in Wissenschaft und Technik, 2*, 203–221.

Mejia, C., & Kajikawa, Y. (2017). Bibliometric analysis of social robotics research: Identifying research trends and knowledgebase. *Applied Sciences, 7*(12), 1–17.

Miroshnichenko, A. (2018). AI to bypass creativity. Will robots replace journalists? (The answer is "yes"). *Information, 9*(7), 183–203.

Mogg, T. (2019, March 4). Jibo the social robot is about to become an expensive ornament. *Digital Trends*. https://www.digitaltrends.com/cool-tech/jibo-the-social-robot-is-about-to-become-an-expensive-ornament/

Monahan, T. (2008). Marketing the beast: Left behind and the apocalypse industry. *Media, Culture & Society, 30*(6), 813–830.

Moravec, H. (1988). *Mind children: The future of robot and human intelligence*. Harvard University Press.

Mori, M. (2012, June). The uncanny valley. *IEEE Robots & Automation Magazine, 19*(2), 98–100. http://ieeexplore.ieee.org/xpl/tocresult.jsp?isnumber=6213218&punumber=100

Mubin, O., Wadibhasme, K., Jordan, P., & Obaid, M. (2019). Reflecting on the presence of science fiction robots in computing literature. *ACM Transactions on Human-Robot Interaction (THRI), 8*(1), 1–25.

Mutlu, B., & Forlizzi, J. (2008, March). Robots in organizations: The role of workflow, social, and environmental factors in human-robot interaction. In *2008 3rd ACM/IEEE International Conference on Human-Robot Interaction (HRI)* (pp. 287–294). IEEE Press.

Nasir, L. (2020, July 12). *Ice Cube declares hate for robots... Is he afraid of the Singularity? The Things,* https://www.thethings.com/ice-cube-declares-hate-for-robots-is-he-afraid-of-the-singularity/

Nassauer, S. (2020, November 2). Walmart scraps plan to have robots scan shelves. *Wall Street Journal.* https://www.wsj.com/articles/walmart-shelves-plan-to-have-robots-scan-shelves-11604345341

Natale, S. (2021, May 10). Artificial intelligence and gullible humans. *IAI, 96.* https://iai.tv/articles/ai-lies-and-deception-auid-1805

Nicholson, I. (2011). "Shocking" masculinity: Stanley Milgram, "obedience to authority", and the "crisis of manhood" in Cold War America. *Isis, 102*(2), 238–268.

Nijholt, A. (2019, January). Stand-up comedy and humor by robots. In *Proceedings Sixteenth International Symposium on Social Communication* (Vol. 1, pp. 228–234).

Njuguna, J. (2021). Constructions of moral values in reader comments of the Samantha sex robot discourse in East African newspapers. *Journal of Religion, Media and Digital Culture, 10*(3), 382–403.

Norman, D. (1992). *Turn signals are the facial expressions of automobiles.* Diversion Books.

Nourbakhsh, I. R. (2015). The coming robot dystopia. *Foreign Affairs, 94*(4), 23–28.

Olaronke, I., Rhoda, I., Gambo, I., Oluwaseun, O., & Janet, O. (2020). A systematic review of swarm robots. *Current Journal of Applied Science and Technology, 39,* 79–97.

Oravec, J. A. (1996). *Virtual individuals, virtual groups: Human dimensions of groupware and computer networking.* Cambridge University Press.

Oravec, J. A. (2004). Examining the examined career: Diana Forsythe as ethnographer and participant in computing research. *Science and Public Policy, 31*(2), 159–163.

Oravec, J. A. (2019). The "dark side" of academics? Emerging issues in the gaming and manipulation of metrics in higher education. *The Review of Higher Education, 42*(3), 859–877.

Oravec, J. A., & Travis, L. (1992). If we could do it over, we'd... Learning from less-than-successful expert system projects. *Journal of Systems and Software, 19*(2), 113–122.

Quadflieg, S., Ul-Haq, I., & Mavridis, N. (2016). Now you feel it, now you don't: How observing human-robot interactions and human-human interactions can make you feel eerie. *Interaction Studies, 17*(2), 211–247.

Pace, J. (2017). Exchange relations on the dark web. *Critical Studies in Media Communication, 34*(1), 1–13. https://doi.org/10.1080/15295036.2016.1243249

Parsa, C. (2019). *Artificial intelligence dangers to humanity: AI, U. S, China, Big Tech, Facial Recognition, Drones, Smart Phones, IoT, 5G, Robotics, Cybernetics, and Bio-Digital Social Program*. The AI Organization.

Pasquale, F. (2020). *New laws of robotics: Defending human expertise in the age of AI*. Belknap Press.

Pesce, N. L. (2021, July 26). *He shoots, he scares! Toyota's basketball robot steals the show at the Tokyo Olympics*. Marketwatch.com. https://www.marketwatch.com/story/he-shoots-he-scares-toyotas-basketball-robot-steals-the-show-at-the-tokyo-olympics-11627313853

Pope Francis, November Prayer Intention, Vatican, 5 November 2020.

Postnikoff, B. (2020). *Robot social engineering* (Master's thesis). University of Waterloo. http://hdl.handle.net/10012/16030

Pymnts. (2021, June 29). SoftBank puts 'Pepper' robot on hold as bot biz does some soul-searching. https://www.pymnts.com/news/artificial-intelligence/2021/softbank-puts-pepper-robot-on-hold-as-bot-biz-does-soul-searching/

Ramos, A. C. S., Contreras, V., Santos, A., Aguillon, C., Garcia, N., Rodriguez, J. D., Vazquez, I. A., & Strait, M. K. (2018). A preliminary study of the effects of racialization and humanness on the verbal abuse of female-gendered robots. *ACM/IEEE International Conference on Human-Robot Interaction*, 227–228. https://doi.org/10.1145/3173386.3177075

Rasib, M., Butt, M. A., Khalid, S., Abid, S., Raiz, F., Jabbar, S., & Han, K. (2021). Are self-driving vehicles ready to launch? An insight into steering control in autonomous self-driving vehicles. *Mathematical Problems in Engineering, 2021*. https://doi.org/10.1155/2021/6639169

Ray, A. (2018). *Compassionate artificial intelligence: Frameworks and algorithms*. Compassionate AI Lab (An Imprint of Inner Light Publishers). https://amitray.com/compassionate-artificial-intelligence-scopes-and-challenges/

Reader, J. (2021). From new materialism to the postdigital: Religious responses to environment and technology. *Postdigital Science and Education, 3*(2), 546–565.

Reedy, C. (2017, October 5). Kurzweil claims that the Singularity will happen by 2045. *Futurism*. https://futurism.com/kurzweil-claims-that-the-singularity-will-happen-by-2045

Ryland, H. (2021). Could you hate a robot? And does it matter if you could? *AI & Society, 36*, 637–649.

Ryznar, M. (2018). Robot love. *Seton Hall Law Review, 49*, 353–374.

Sachgau, O. (2017, September 18). Robots endangering workers drives German startup to airbags. *Industry Week*. https://www.industryweek.com/operations/article/22024196/robots-endangering-workers-drives-german-startup-to-airbags

Sanoubari, E., Seo, S. H., Garcha, D., Young, J. E., & Loureiro-Rodríguez, V. (2019, March). Good robot design or Machiavellian? An in-the-wild robot leveraging minimal knowledge of passersby's culture. In *2019 14th ACM/IEEE International Conference on Human-Robot Interaction (HRI)* (pp. 382–391). IEEE Press.

Shah, S. (2021, July 21). A robot collision sparked a fire at the UK's top online grocer: Ocado was forced to evacuate the facility and cancel orders. *Engadget*. https://www.engadget.com/ocado-robot-collision-fire-114027027.html

Share, P., & Pender, J. (2018). Preparing for a robot future? Social professions, social robotics and the challenges ahead. *Irish Journal of Applied Social Studies, 18*(1), 4–18.

Shelley, M. (1818). *Frankenstein: A modern Prometheus*. Independently published.

Simoens, P., Dragone, M., & Saffiotti, A. (2018). The Internet of Robotic Things: A review of concept, added value and applications. *International Journal of Advanced Robotic Systems, 15*(1), https://doi.org/10.1177/1729881418759424

Skal, D. J. (1998). *Screams of reason: Mad science and modern culture*. W. W. Norton.

Slaughter, R. A. (2021, forthcoming). The IT Revo Part 4: Transcending the Matrix. *Futures*. https://doi.org/10.1016/j.futures.2021.102869

Smith, A. (2018, August 29). Franken-algorithms: The deadly consequences of unpredictable code. *The Guardian*. https://www.theguardian.com/technology/2018/aug/29/coding-algorithms-frankenalgos-program-danger

Solon, O. (2017, September 28). Deus ex machina: Former Google engineer is developing an AI god. *The Guardian*. https://www.theguardian.com/technology/2017/sep/28/artificial-intelligence-god-anthony-levandowski

Sone, Y. (2016). *Japanese robot culture*. Springer.

Song, Y. S. (2021). Religious AI as an option to the risks of superintelligence: A Protestant theological perspective. *Theology and Science, 19*(1), 65–78.

Sophia. (2020, March 31). *We need creativity*. https://twitter.com/RealSophiaRobot/status/1245092521014964225?r-most-famous-robot%2F

Spatola, N., & Urbanska, K. (2020). God-like robots: The semantic overlap between representation of divine and artificial entities. *AI & Society, 35*(2), 329–341.

Spielberg, S. (2018). *Ready player one*. Warner Bros USA.

Stahl, W. A. (1995). Venerating the black box: Magic in media discourse on technology. *Science, Technology, & Human Values, 20*(2), 234–258.

Steinert, S. (2016). Taking stock of extension theory of technology. *Philosophy & Technology, 29*(1), 61–78.

Stephenson, N. (2014). *Snow crash*. Bragelonne.

Stilgoe, J. (2018). Machine learning, social learning and the governance of self-driving cars. *Social Studies of Science, 48*(1), 25–56.

Sudharsan, R. R., & Deny, J. (2021). Brain–computer interface using electroencephalographic signals for the Internet of Robotic Things. In R. Anandan, G. Suseendran, S. Balamurugan, A. Mishra, & D. Balaganesh (Eds.), *Human communication technology: Internet of Robotic Things and ubiquitous computing* (pp. 27–53). Wiley. https://doi.org/10.1002/9781119752165.ch2

Szollosy, M. (2017). Freud, Frankenstein and our fear of robots: Projection in our cultural perception of technology. *AI & Society, 32*(3), 433–439.

Terzian, S. G. (2009). The 1939–1940 New York World's Fair and the transformation of the American science extracurriculum. *Science Education, 93*(5), 892–914.

Tonkin, S. (2021, December 3). Meet Ameca. *Daily Mail*. https://www.dailymail.co.uk/sciencetech/article-10270925/Worlds-advanced-humanoid-robot-unveiled-UK-lab.html?fbclid=IwAR1Ta9EidYxN8Vhah4ijjbBStko-MQ3Fbh Lai1j0EFGq7kogx28DPSPAx6A

Trovato, G., Lucho, C., Huerta-Mercado, A., & Cuellar, F. (2018, March). Design strategies for representing the divine in robots. In *Companion of the 2018 ACM/IEEE International Conference on Human-Robot Interaction* (pp. 29–35). https://doi.org/10.1145/3173386.3173388

Turing, A. (1950). Computing machines and intelligence. *Mind, 49*, 433–460.

Ullman, D., & Malle, B. F. (2018, March). What does it mean to trust a robot? Steps toward a multidimensional measure of trust. In *Companion of the 2018 ACM/IEEE International Conference on Human-robot Interaction* (pp. 263–264). https://dl.acm.org/doi/abs/10.1145/3173386.3176991

Ullrich, D., Butz, A., & Diefenbach, S. (2021). The development of overtrust: An empirical simulation and psychological analysis in the context of human-robot interaction. *Frontiers in Robotics and A, 1*, 8. https://doi.org/10.3389/frobt.2021.554578

Van Aerschot, L., & Parviainen, J. (2020). Robots responding to care needs? A multitasking care robot pursued for 25 years, available products offer simple entertainment and instrumental assistance. *Ethics and Information Technology, 22*, 247–256.

van Wynsberghe, A. (2021). Responsible robotics and responsibility attribution. In J. von Braun, M. S. Archer, G. M. Reichberg, & M. Sánchez Sorondo (Eds.), *Robotics, AI, and humanity science, ethics, and policy* (pp. 239–249). Springer International Publishing.

Vanderelst, D., & Winfield, A. (2018, December). The dark side of ethical robots. In *Proceedings of the 2018 AAAI/ACM Conference on AI, Ethics, and Society* (pp. 317–322). ACM Press.

Verbeek, P. (2011). *Moralizing technology: Understanding and designing the morality of things.* University of Chicago Press.

Vilk, J., & Fitter, N. T. (2020, March). Comedians in cafes getting data: Evaluating timing and adaptivity in real-world robot comedy performance. In *Proceedings of the 2020 ACM/IEEE International Conference on Human-Robot Interaction* (pp. 223–231).

Voss, L. (2021). Showing off robots: In/animacy attributions in robotics demonstrations, science communication, and marketing. In *More than machines?* (pp. 75–104). transcript-Verlag.

Walter, E. (2016). RIT team gets ready for BattleBots show in LA. *Democrat and Chronicle*. https://www.democratandchronicle.com/story/money/2016/04/08/rit-students-engage-battlebots-robotic-combat/82481746/

Waqar, B. (2021, April 3). People have started throwing eggs at Google's autonomous vans. *Wonderful Engineering*. https://wonderfulengineering.com/people-have-starting-throwing-eggs-at-googles-autonomous-vans/

Watt, M. C., Maitland, R. A., & Gallagher, C. E. (2017). A case of the "heeby jeebies": An examination of intuitive judgements of "creepiness." *Canadian Journal of Behavioural Science/revue Canadienne Des Sciences Du Comportement, 49*(1), 58.

Webster, S. (2021, November 12). American companies are adding more robots in its workforce to keep up with demand. *TechTimes*. https://www.techtimes.co//////m/articles/267921/20211112/american-companies-adding-more-robots-workforce-keep-up-demand.htm

Weinberg, G., Bretan, M., Hoffman, G., & Driscoll, S. (2020). *Robotic musicianship: Embodied artificial creativity and mechatronic musical expression* (Vol. 8). Springer Nature.

Weng, Y. H., Chen, C. H., & Sun, C. T. (2009). Toward the human–robot co-existence society: On safety intelligence for next generation robots. *International Journal of Social Robotics, 1*(4), 267–282. https://doi.org/10.1007/s12369-009-0019-1

Wight, J. K. (2020). The battle for the robot soul. *Philosophy Now, 139*, 16–19. https://philosophynow.org/issues/139/The_Battle_for_the_Robot_Soul

Winkle, K., Melsión, G. I., McMillan, D., & Leite, I. (2021). Boosting robot credibility and challenging gender norms in responding to abusive behaviour: A case for feminist robots. *ACM/IEEE International Conference on Human-Robot Interaction*, 29–37. https://doi.org/10.1145/3434074.3446910

Winner, L. (1978). *Autonomous technology: Technics-out-of-control as a theme in political thought.* MIT Press.

Wittes, B., & Blum, G. (2015). *The future of violence: Robots and germs, hackers and drones-Confronting a new age of threat*. Basic Books.

Wolfert, P., Deschuyteneer, J., Oetringer, D., Robinson, N., & Belpaeme, T. (2020, March). Security risks of social robots used to persuade and manipulate: A proof of concept study. In *Companion of the 2020 ACM/IEEE International Conference on Human-Robot Interaction* (pp. 523–525). IEEE Press.

Wootson, C. (2017, October 29). Saudi Arabia, which denies women equal rights, makes a robot a citizen. *Washington Post*. https://www.washingtonpost.com/news/innovations/wp/2017/10/29/saudi-arabia-which-denies-women-equal-rights-makes-a-robot-a-citizen/

Woźniak, P. W., Karolus, J., Lang, F., Eckerth, C., Schöning, J., Rogers, Y., & Niess, J. (2021, May). Creepy technology: What is it and how do you measure it? In *Proceedings of the 2021 CHI Conference on Human Factors in Computing Systems* (pp. 1–13).

Yam, K. C., Bigman, Y. E., Tang, P. M., Ilies, R., De Cremer, D., Soh, H., & Gray, K. (2020). Robots at work: People prefer—And forgive—Service robots with perceived feelings. *Journal of Applied Psychology, 106*(10), 1557–1572. https://doi.org/10.1037/apl0000834

Young, A. G., Majchrzak, A., & Kane, G. C. (2021). Organizing workers and machine learning tools for a less oppressive workplace. *International Journal of Information Management, 59*, 102353.

Zaman, B., Van Mechelen, M., & Bleumers, L. (2018, June). When toys come to life: Considering the Internet of Toys from an animistic design perspective. In *Proceedings of the 17th ACM Conference on Interaction Design and Children* (pp. 170–180).

Zorc, J. J., Chamberlain, J. M., & Bajaj, L. (2019). Machine learning at the clinical bedside—the ghost in the machine. *Journal of the American Medical Association (JAMA) Pediatrics, 173*(7), 622–624.

Zweig, C., & Abrams, J. (1991). *Meeting the shadow: The hidden power of the dark side of human nature*. TarcherPerigee.

CHAPTER 4

Love, Sex, and Robots: Technological Shaping of Intimate Relationships

Sex-related robots and AI-enhanced entities are proliferating as consumer objects worldwide, exhibiting various aspects of human-to-robot and human-to-AI intimate interactions (Björkas & Larsson, 2021; Lee, 2017; Liu, 2021). An upcoming chapter centers on the issue of the abuse by humans of robots, and this entire book focuses on darker and less positive aspects of technological initiatives. Thus, this chapter's focus on the often-unsettling controversies involving sex robots and other forms of AI-enabled sexual activity is not surprising. Sex robots are playing critical roles in the ecology of robot and AI-enhanced entities. The question of whether robots "outclass" humans (discussed in previous chapters) is construed in a different manner when the robot involved is one's own specially-designed sexual partner or even spouse. Many of the kinds of contact individuals have with robots are indeed disempowering (such as with surveillance-related robots); however, the commodification of various aspects of sex robot individuation discussed in this chapter can serve to increase the perception of mastery and control that would otherwise not be available to many people. One specific instance here is how the sex robot users who design a personalized robot to their specifications (often with celebrity figures or personal acquaintances as themes) may recapture some of the perceived mastery that they lost during automation-related employment crises or when they were detained by a security

© The Author(s), under exclusive license to Springer Nature Switzerland AG 2022
J. A. Oravec, *Good Robot, Bad Robot*, Social and Cultural Studies of Robots and AI, https://doi.org/10.1007/978-3-031-14013-6_4

robot. Whether this perceived empowerment ultimately plays strong roles in bolstering the users' personal and social wellbeing is yet another matter.

A new assortment of expressions has been coined to capture the broadening spectrum of what is going on with sex, robots, and AI. Human interaction with these sexualized technological entities is sometimes known as "erobotics" (Dubé & Anctil, 2020), though the language of technosex has yet to stabilize. The term used for child sex robots is often "paedobots" (Cox-George & Bewley, 2018), but other labels are developing. Sex robots have been labeled as "synthetic companions" (Zara et al., 2021), "artificial companions" (Coeckelbergh, 2010), and "sexual appliances" as well as "pornbots" (as discussed in a following section). Such terms as "digisexuals" or "robosexuals" emerged in the past decade, identifying individuals who associate strongly with human–robot interactions (McArthur & Twist, 2017). The love and appreciation of robots can include other forms than human sexual expression, with the term "lovotics" sometimes used as an umbrella term to include these deep sensitivities (Cheok et al., 2017; Levy, 2007). Since sex robots have become available on the market, a great deal of attention has been directed to their prospects in popular and academic discourse; these narratives add to the symbolic and thematic inheritances that are generally associated with robots and other AI-related entities.

The development and proliferation of robots that are designed to enchant, tease, excite, and distract humans are not a good way to foster seriousness and vigilance in using robots and other AI-related entities, approaches that can indeed be life-saving in some circumstances. The notions of "creepiness" or the "heeby jeebies" often emerge in discussions of sex robots, which may be a factor in the current and upcoming moral panics related to the entities (Gersen, 2019; Watt et al., 2017). As discussed in the opening chapters, creepiness is often linked closely with privacy invasions, such as the kinds that readily appear with sex robots and related AI entities. Individuals who are compliant enough to buy the newest sex robot and have aspects of their intimate lives subsequently recorded and shaped by AI systems may be malleable enough to work for organizations that are designed for the efficient operation of robot hardware, as well as live in regions that are controlled by military and police robots. Sex robots that are designed to have connections with public computer networks in effect transform sex to be a public act that can be utilized for micro-level control of individuals, and possibly even the prediction of birth rates in various locales:

As people increasingly use humanoid robots as sexual partners, particularly in developed countries where individuals can afford expensive sexbots, the birth rate of developed countries will surely fall from the current 1.7 in the US, 1.6 in Europe, and 1.4 in Japan. (McBride, 2021, p. 1)

Considering sex robots in the context of this book's "dark side of technology" initiative is more than a trivial or diversionary exercise, especially when child sex issues are involved (Chatterjee, 2020). Sex robot technology has the potential to affect the larger range of human–robot–AI interactions in the home, workplace, and community. For example, the growing international investments in the sex robot arena can serve to drive the direction of the personal robot industry, having financial as well as design-level impacts on robotics and AI as a whole. Sex robots are readily incorporated in "dark patterns" schemes, with distracted and excited users potentially interacting with the entities in ways that are not in their best economic or social self-interest (Lacey & Caudwell, 2019). The normalization of intimate data collection as a part of sex robot and related platform processes can also have especially troubling implications for privacy extending beyond the collection of personal data through health wearables (Lupton, 2015; Oravec, 2020, 2022). Also troubling are the growing linkages between sexual activity with robots and some kind of spiritual communion with the Singularity, which is discussed in a section to come. The notion that sex and marriage with robots have a spiritual and even other-worldly component opens many concerns about robots and AI being construed as somehow "outclassing" humans and exhibiting a kind of essential superiority.

Social media interactions concerning sex robots have had powerful impacts on the technologies involved. Consider the following description of RealDoll Corporation, a "Westworld-style sex robot factory [that] promises 'exciting' new dolls in 2022":

> RealDoll has manufactured a number of realistic dolls with Olivia already having three variations. The company even allows customers to build their own doll using different heads, bodies, and makeup style.
>
> Their Instagram account that boasts over 46k followers has posted everything from barely completed dolls and robots to staff painting dildos. (Card, 2021)

Wade-Palmer (2021) describes the offer of some current and potential human customers of RealDoll to bankroll new kinds of sex robot technologies, with the request to add more "artificial intelligence" to the company's robotic offerings. The addition of various personality nuances to the robots has blurred the line between mere sex appliance and personal companion; for example, some of the RealDoll robots reportedly remember individuals' birthdays and quote Shakespeare (Tosin, 2017).

Some of the important issues concerning sex robots may indeed have tangents with other kinds of AI-enhanced entities (such as autonomous vehicles), as individuals acquire skills, expectations, and habits pertaining to AI technology from their sexual and commercial interactions with sex robots and other AI sexual vehicles, such as chatbots. Sex robots may be some of the most complex and nuanced robots that many individuals will encounter in their everyday interactions, at least in the near future; the feelings and moral tone they experience in relation to sex robots have implications for their other technological and social interactions. For individuals who do not have extensive exposure to the programming and maintenance of manufacturing and service robots, observing or using a sex robot can inform them about these AI-enhanced technologies, which can be unfortunate if their characterizations of the sex robot's operations are deficient or intentionally skewed by marketers, advertisers, and distributors. The danger of the development of assortments of dark patterns in relation to sex robots is of special significance: sex robots can be readily programmed in ways that manipulate the users' perceptions, elicit inappropriate information and permissions, and distort their notions of the situation while they are unable (or less able) to give rational consent (Luguri & Strahilevitz, 2021).

Such forms of manipulation as blackmail and gaslighting are also feasible with sex robots, making enhanced security for the entities an imperative. Sexually-themed forms of blackmail have been part of human relations for centuries and are now considerable factors in the information-intensive context of sex robots. UK professor Rob Brooks is quoted as stating "If you're having simulated relationships with machines you would want it to be private because the capacity for blackmail is high, but obviously manipulation is high too" (Nazaroff, 2021, para. 5). Since a good deal of human-to-human sexual activity has the themes of subtle manipulations and distortions, the potential for opportunist and manipulative robotic design in the sex robot arena looms large. For example, during sexual arousal, specific activities or verbal commands on the part of users could be compelled to purchase particular products, produce certain physical results, or release intimate kinds of information without the users' lucid and fully cogent consent. The potential for forms of sexually-themed "gaslighting" is substantial (strategies that take advantage of users' anxieties and misapprehensions); gaslighting can be used as part of social engineering to debilitate and control people as well as efforts by corporations to make consumers acquire updated technologies, under the assumption that the current version is malfunctioning. The term "gaslighting" is linked with the 1938 play *Gas Light* (as well as a 1944 film entitled *Gaslight* starring Ingrid Bergman) in which various environmental and psychological methods were used to distort a young woman's sense of reality and feelings of competence.

Some Historical Insights on Technosex

Sex robots have generated a great deal of discourse, despite the fact that the consumer market for these robots is just developing. They currently have an enlarged and often exaggerated place in relation to technosex and the sexual imaginary in comparison with their numbers and overall usage, just as the overall impacts of some military and police robots can extend beyond their immediate reach; sex robots are still relatively sparse in number, although this can change as commercial efforts continue. They have indeed generated extensive publicity, especially in the tabloid journalism of the US and UK. For example, the news that "First lesbian sex robots may be coming soon as sexbots start being programmed to be gay" (Salas-Rodriguez, 2021) was distributed to Internet audiences via the US version of *The Sun*.

Applications of robotics and AI technology can expand the capabilities of sexual expression in many ways, with special emphasis on how the expressions are mass produced and disseminated. Sex dolls, some with seemingly animate features, have been a part of human intimate expression for centuries (Cheok & Zhang, 2019). Early uses of mainframe computers in corporate and university settings included the production and sharing of sexually-oriented and pornographic files of various sorts, including narratives and images (Paasonen, 2011). Through the 1960s, the development and sharing of these files were generally more individualized and less commercialized, with the designers' personal creativity and initiative more a factor than marketable production values. As the computing industries developed through the following decades, the economic benefits of developing and disseminating sexually-themed products on a commercial basis became clearer, although the non-commercial production of sexual content also continued (Coopersmith, 2006). Levy (2007) relates how "... recently, sex has led some of the most important technological developments within the consumer-electronics boom in sales of the videocassette recorder (porn videos), then the DVD (more porn), and, of course, the Internet (yet more porn ..." (p. 289). AI-enhanced chatbots as well as human-powered erotic chatrooms also became lucrative parts of the Internet-centered technosex arena in the past several decades, though forms of erotic technologically-mediated chat extended years before that (Nagy, 2021).

Technosex-related public discourse and critiques often tend to be selective, focusing not on overall industry directions but creating social issues concerning particular applications or services. The considerable attention to Craigslist's personal ads (Oravec, 2014) and sexting concerns (Oravec, 2012) began to shape the discourse around technosex topics in the past decade, also bringing in law enforcement. In subsequent years, focus shifted to social media expansions and the presence of sex workers in such venues as the OnlyFans platform (O'Brien, 2021). With the scourge of AIDS in the 1980s and the recent COVID-19 pandemic affecting everyday social interactions, controversies involving technosex have often taken on a different character (Effoduh, 2021; Wolf, 2012). The avoiding of some kinds of interpersonal contact (or "social distancing") has been construed by any commentators and legal officials as having a positive moral purpose (stopping the spread of diseases), often directing societal attention away from negative aspects of the technologies, however severe in their implications.

The technosex industry has produced many kinds of physical and virtual items through the past decades (with revenues in the billions of dollars), but sex robots have only recently emerged as a popularly-available commercial product. The company TrueCompanion (US, New Jersey) developed Roxxxy in 2010, which is often recognized as one of the first commercial sex dolls to utilize AI approaches (Cotton, 2018). A large assortment of AI-enhanced sex robots continued to be produced, with a wide variety of claims about their abilities, medical benefits, and their levels of consciousness. Such claims cannot yet be vetted and regulated by governmental agencies, since sex robots have not yet been declared as medical devices in the US and many other nations; a wide spectrum of other kinds of computerized health devices is also in this nebulous category, including health wearables (Oravec, 2020, 2022). Increases in development and proliferation of sex robot initiatives may seem to be inevitable in many national contexts, with the assumption that sex robot-related technical developments proceed in the rapid manner as those of various other technosex entities (such as online dating platforms, sex-themed chatbots, and sexting). This speed will give public policy analysts and the medical community little time to weigh in on the personal safety and social appropriateness of the entities. Many of the applications discussed in this chapter raise obvious legal issues. Child sex robots, for instance, are generally considered as entities that can be constrained legally, though the commercially-available sex robot architectures can enable users to produce robots with child sex themes themselves. Some researchers have suggested, however, that child sex robot utilization may actually play a role in mitigating the negative impacts of certain kinds of pedophilia and may thus eventually become a part of therapeutic measures, which may increase the commercial availability of child sex robots (Brown & Shelling, 2019; Maras & Shapiro, 2017).

Positive characterizations of sex robots and defenses of their utilization are increasing in variety, including the habits that quantitative assessments of exercise-related behavior by the robot can provide: "a robot, for example, can encourage proper exercise and discipline by giving positive feedback to its users" (Peeters & Haselager, 2021). Life-enhancing aspects of sex robot applications can certainly be projected (as in Bendel, 2015; Jecker, 2021, which focus on the use of robots for medical care and individuals with disabilities). Döring and Pöschl (2018) argues that sexual products have the potential "to improve sexual well-being in various populations" (para. 1). However, these potentially encouraging aspects are not the theme of this book. Some unsettling descriptions of sex robots have raised alarms, such as the following excerpt from an article in *The Australian* about a sex robot produced by the US corporation TrueCompanion: "The first thing I noticed about Roxxxy was not her improbable proportions but the look frozen on her face, somewhere between pleasant surprise and utter horror" (Brooks, 2021b, para. 1). The similarities of such robotic expressions with human expressions of anguish and pain present unsettling parallels. Sexual contact is indeed often a part of the modalities of war and the conquering of people, of breaking the spirits of the oppressed; in comparable manners, forms of pedophilia, misogyny, and racism can be involved in technosex production.

Sexual and Marriage Unions with Robots: The Singularity and Sexual Imagery

The melding and communion of humans and robots at a spiritual and even mystical level is taking a variety of forms and can add additional complexities to sex robot issues for some individuals as they discuss and compare themselves with their robotic partners. Sex robots are just one of the ways in which individuals can seemingly connect with robots and other AI-enhanced entities, enabling humans to share in some ways in the supposed intellectual and spiritual superiority of AI. The legacy of robotics in science fiction and folklore as well as the near deification of AI in many Western societies has extended the potential influences of sex robots and related AI technosex entities, associating them with power and the future rather than the mundane everyday quality of often-disappointing human existence. The human–robot sexual and marriage unions described in this chapter (as well as the fictionalized accounts of these sexual unions) are part of the adjustments of the social order that serve to incorporate robots and AI. Humans who interact with robots on a sexual level may become more open to integrating robots in the workplace and community, and perhaps the battlefield and police station as well; sexual interaction with robots makes for a more compliant and less questioning consumer of robotics and AI. Such romantic acts can also, for some individuals, reflect the transformation of robotics and AI into spiritual, magical, and even religious realms. For instance, philosopher Amit Ray (2018) stated that "Humans are limited in the attention, kindness and compassion that they can expend to others, but AI based compassionate robots can channel virtually unlimited resources into building compassionate relationships in the society," projecting that robots can actually be superior to humans and outclass them in these capacities.

An assortment of transformational unions that are emerging as many humans (reflected in sexual acts and marriage) seek to take their places in the singularity. The singularity involves the cascading growth of robotic and AI intelligence to the point at which it exceeds human intelligence and takes on new dimensions, possibly out of the control or even knowledge of the humans involved (Eden et al., 2012). Humans who do not want to be left behind as the singularity emerges may want to use romance and even marriage as a way to maintain a union with the AI-enhanced entities. The association of "romance" with sex robots may seem off-putting, but relates to some of the larger issues concerning human interactions with technology. Maldonato et al. (2021) write of the "sex appeal" of algorithms and the inorganic. Lunceford (2018) portrays the union of human and robot in romantic terms: "the singularity could not be completely non-organic; it must take place in the form of a cyborg, wedding the digital to the biological" (p. 221). The capabilities of designing specially-tailored humanoid characteristics in sex robots can magnify the dimensions of sexual expression, expanding it beyond human limitations:

> ... a new materialist, sex positive, queer perspective, the article proposes a realignment of sex robotics towards post gender forms and materialities. Liberating sex robots from the obligation to imitate the human body as faithfully as possible, may also serve as an antidote to the eeriness and revulsion many people experience when having to deal with humanoid – but not quite human – replicants. (Kubes, 2019)

To what extent are humans attempting to fuse with the supposedly-superior robotic and AI entity through sexuality? Analyzing the narratives related to human–robot interaction (as described in Mainenti, 2020) may provide some clues, and movie and literature themes may provide directions as well; the marketing efforts of advertisers who wish to shape how humans perceive robots may also have considerable influence.

Marriage with robots and other AI-enhanced entities has become a related theme, integrating the notion of spiritual union from a different angle (Yanke, 2021). Science fiction has explored these issues for decades: Isaac Asimov's (1983) novel *The Robots of Dawn* examines related human–robot sexual and partnership themes, emphasizing their complexities and their importance to the emerging social order. The drama associated with assimilation struggles (the robot becoming part of human households) is portrayed in *Bicentennial Man*, starring the late Robin Williams (Inniss, 2001). Aifric Campbell's (2021) novel *The Love Makers* explores the intricacies of the near future of sex robots from the angle of two women traveling companions.

An example of robot marriage complications is *Ride, Sally, Ride* by Douglas Wilson (2021) which explores the charging of a man with murder for destroying his neighbor's sex robot (which was claimed to be a legal spouse):

It's two decades in the future, and a Christian college student named Ace Hartwick has just destroyed his neighbor's so-called "wife"—actually a sexbot named Sally—in a trash compactor. Soon, Ace will be on trial for murder. (https://canonpress.com/products/ride-sally-ride-a-novel/)

In the 2021 US movie *I'm Your Man*, the prospects for domestic and sexual bliss with a humanoid robot were portrayed:

> … for three weeks, she [the protagonist] must live with Tom… a humanoid robot that has been precision-tooled to be her perfect companion… Tom has an unblinking LED-blue gaze and a database full of cornball compliments – "Your eyes are like two mountain lakes," he informs her, with the same tone of polite interest that he uses when calculating the optimal angle of safety for her car seat. But Tom is programmed to learn from her responses. And even though, or perhaps because, Alma shuts him in a cupboard along with her ironing board and rejects his offer of a petal-strewn scented bath, he starts to more closely resemble the kind of man she could live with, or even love. (Ide, 2021, para. 2–3)

The lighthearted and sociable treatment of the subject matter in *I'm Your Man* contrasts with the more serious projections of some researchers that psychologists will soon have to deal with "an increasing solipsism stemming from people engaging with machines like sex robots" (Hasse, 2019). Themes of other kinds of sexually-active AI-enhanced devices are proliferating in film and literature. For example, Kermode (2021) discusses the 2021 award winning film *Titane*, in which an autonomous vehicle impregnates a woman.

Currently, robot-human marriages have not yet been legally sanctioned by countries, despite the interest in them by some individuals, as reflected in this narrative: "'Proud Robosexual' Plans to Marry Robot When It's Legal" (2016). Four thousand users have reportedly "married" their robotic companions (designed with holographic technology) with certificates issued by Gatebox Corporation; they have reportedly signed statements to the effect that "I love her [the robot] and see her as a real woman" (*The Nature of Things*, 2021). Whether the signers are seriously coupled with their holographic mates or are engaging in some kind of parody of marriage is uncertain, but these sexually-themed activities are opening up new vistas for human–robot interaction and communion. Nations have taken various stances on sex robot regulation in general (whether or not marriage is involved); for example, "no Australian legislation currently regulates or prohibits sexual intercourse with robots, there are regulations on child-like sex dolls which have been addressed by the Commonwealth, South Australia, and Queensland" (Nield, 2021). The legislators looking into potential sex robot regulations generally require some sorts of cases of specific personal or public health harms before intervening. Some of the emerging robotic-related sexual applications in effect commercialize the capabilities to do some substantially harmful things to oneself and others, with the potential for physical damage as well as psychological trauma (Effoduh, 2021); for example, unless stresses on the hearts of certain human participants are monitored, some physical damage may occur.

Dark Sides of Human–Robot Sexual and Romantic Communion

A wide assortment of spiritual and romantic versions of the human-meets-robot theme has emerged, as evidenced in the past sections. Many other, darker motivations are certainly apparent (and more are emerging), however. The range of expressiveness or embodiment of sex robots has ramifications for how they are utilized (van Grunsven & van Wynsberghe, 2019); misogynistic imageries and themes are common in some sex robot configurations; misogynistic, transphobic, and other pejorative themes are a part of how many sex robots are apparently used and discussed in practice. Headlines such as "Woman cloned into sex doll so realistic you'd need microscope to tell them apart" (Torre, 2021) can indeed generate discourse about humans in relation to robots, but can also exacerbate difficult and often damaging gender issues. Sex robots can include entities with humanoid characteristics based on particular people, including famous people as well as non-celebrity individuals chosen by the consumer. The latter variety are involved in efforts toward the engineering of sexual objects (discussed in the upcoming section on the "new Pygmalion"), which can have larger cultural as well as personal ramifications. For example, Cavanagh (2021) relates how "sex robots could soon be 'cloned' to look like your dead partner thanks to groundbreaking 3-D scans," potentially creating a new market for the compilation of personal characteristics for after-death themed sexuality ("necrorobosexuality") as well as characteristics of unwilling sexual partners such as ex-spouses.

Much of the research literature on sex robots presents the uses of sex robots in clinical ways, without an examination of what it might mean to be making love to a robot. However, in doing so, the "dark side" legacies of robots are often downplayed, omitting some of the overt psychological and cultural meanings of having sex with robots. For instance, the potentials for robotic "revenge sex" in which individuals insert the characteristics of a particular individual into a robot and perform violent or otherwise inappropriate sexually-themed acts with the robot can attract individuals to the sex robot genre who are manifesting psychologically disturbing tendencies. Such revenge sex might be monitored by the organization that controls the platform with which the sex robot is linked, so some complications may occur if blackmail is involved or if the revenge fantasies get acted upon in real life. These dark sides of sex robots are often reinforced rather than diminished by the negative tones of publicity about them, with some individuals positively associating the "naughty" or even disgusting aspects of the robots with human sexuality. The desirability to some people of this supposed disgrace and stigma associated with sexuality in many cultures has been a factor in the growth of other commercial sexual industries as well (Voss, 2015). Some of the dedicated efforts to contain the proliferation of the robots and admonish those who use the robots may thus be counterproductive, reinforcing the power of the robots to be offensive and damaging.

The "backstage" sexual activities that refresh and empower us as human beings are rapidly becoming "frontstage" with these various technological interventions. The "robo-exhibitionist" aspects of sex robots can show how frontstage and backstage aspects can become convoluted, as activities generally intended to be private are placed on a larger public stage though Internet connectivity. The potential for sex robots to influence everyday human relationships is still unclear, but the potentials for the robots to enable external interventions in individuals' intimate expressions are becoming fairly obvious. Many of the sex robots will be connected to the Internet at some point and will provide information about sexual habits and choices, just as some digital health wearables are doing now (Lupton, 2015; Oravec, 2020, 2022). Just by purchasing such an entity, people will be manifesting their sexual identities in some fashion for corporate or governmental analysis. The notion of sex robots has yet to stabilize; the entities can include very simple and crude robots with little personalization of their activities to more sophisticated units with individualized responses based on analysis of biometric data as well as demographics. For instance, some of the sexual engineering initiatives "buttress a 'sex for health' discourse which relies on the collection of intimate data purportedly used to improve current and subsequent teledildonics models" (Flore & Pienaar, 2020, p. 279).

> In the discourse of the artificial person, robotic bodies serve as allegorical displacements of a sexual imaginary: their rigidity and hardness function as a thinly veiled register of an idealized state of arousal, while the focus on relentlessness, insistence, repetition, or mechanical. (Kakoudaki, 2014)

Some insights from robot design initiatives may affect the quality of sex robot interactions. For instance, Székely et al. (2019) conducted research on "how the perception of a robot partner's effort elicits a sense of commitment to human–robot interaction" (p. 234). Sex robots that exhibit "effort" in some way may foster the human user's overall commitment to the sexual interaction.

Taking Sex Robots Seriously

Sex robot themes can make for comic material, and unfortunately, the seriousness of some of the dark side concerns involved can be undermined and trivialized by such comedic treatment. For centuries, crude jokes about gender, racial, and disability themes were a part of everyday interaction in workplaces and community settings; sociologist Dundes (2017) states that we "joke only about what is most serious" (p. 1). Such behavior is often linked to the expression of social unease about a certain issue or theme; in Goffman's (1959) dramaturgical approach, humor can serve as a "catharsis for anxieties" (Steele, 2021, p. 102). The dark sides of sex robots (including child sex implications, misogynistic themes, and sexual addiction issues) provide a chilling backdrop to whatever emerges in terms of the technologies involved, but the discussion of these topics is often treated in a lighthearted and humorous manner that disguises their gravity. Consider the following concerns and activities in Houston, Texas, related to the prospects of sex robot brothel:

> If you've followed our news coverage the past few weeks, you have heard about robot brothels. In fact, it made national news, including on ABC's *The View*:

> "Religious groups in Houston are protesting the opening of a brothel featuring sex robots," co-host Whoopi Goldberg announced to laughter from the audience. And she added: "Why're they worried about it?"
>
> While Goldberg and others on the show laughed it off, the matter is serious for some here in Houston. Mayor Sylvester Turner moved swiftly and presented an ordinance to City Council that expands the meaning of adult arcades to include "anthropomorphic devices" – in other words, humanlike sex dolls or... robots.
>
> It passed unanimously... (Martin, 2018)

In some ways, the Houston religious group reactions can be considered in terms of "moral panics" against the technology (Brooks, 2021a). A Swiss sex robot brothel also received a great deal of press attention (Patient, 2019) as well as related comic treatment in social media. Sex robots as commodities and as part of brothel-style organizations are generating societal discourse, much of which is linked to their potential proliferation and eventual regulation. Nevett (2018) describes the protests of sex workers against robot sex brothels in the UK, some of which were sufficient to trigger a locational change of one of the establishments. However, DiTecco (2020) found that "sex workers, sex work clients, and sex robot/doll users were found to be underrepresented in these conversations" (p. 1), which fails to ground this discourse with the realities of the sex industry.

The expressions of concern from medical experts, public health leaders, and educators (described in the sections below) are unfortunately much fewer than are warranted by the early-stage developments involved. Occasional stories in newspapers and social media have raised alarms, such as "New report finds no evidence that having sex with robots is healthy" in the *Washington Post* (Guarino, 2018), which counters some of the expressions of support and positive press sex robots have attracted. Consider the following popular culture description of a sex robot, circa 2021:

> A woman has been cloned into a sex doll so advanced you'd need a microscope to tell the difference. The silicone lover comes with matching toe hairs and details on the model's skin as fine as 0.03mm.
>
> Cloud Climax is selling the mail-order 'Mao Clone' for £6,299 in the UK.
>
> It is the latest in its EX Doll Clone Series that uses 3D photography to 'clone' women into hyper realistic sex robots.

Photos show the doll emerging from a chamber reminiscent of scenes from sci-fi film *Blade Runner* while others reveal her part-made like in *Ex Machina*.

Each Mao model takes between 8-12 weeks to produce and comes with an implanted hair option for an extra £399... [Using] industrial-grade self-developed scanning system for collecting Mao's biological data... details as small as 0.03mm on the skin are replicated. (Torre, 2021, para. 4)

Positive press about these robotic developments can be easily located, as well as criticisms and organized opposition (the latter discussed in an upcoming section). For example, the International Congress on Love and Sex with Robots met in 2018 to organize support and exchange research on sex robots. As outlined in this chapter, some sex robot advocates have presented strong and articulate defenses of the use of sex robots in particular circumstances: "I argue that the right to sexual satisfaction of severely physically and mentally disabled people and elderly people who suffer from neurodegenerative diseases can be fulfilled by deploying sex robots; this would enable us to satisfy the sexual needs of many who cannot provide for their own sexual satisfaction; without at the same time violating anybody's right to sexual self-determination" (Di Nucci, 2016, p. 73). Serious critiques abound, however, as discussed in the next section.

Critiques of Sex Robots

As sex robots have become more widely available in society, their negative assessments have expanded; a number of people who oppose sex robots have developed compelling arguments and searing criticisms. Kathleen Richardson's 2015 paper, "The Asymmetrical 'Relationship': Parallels Between Prostitution and the Development of Sex Robots" served to stimulate widespread discussion of sex robot concerns as well as inspire further research and commentary (Klein & Lin, 2018). Expressing comparable levels of alarm, Galaitsi et al. (2019) labeled the growing concerns about sex robot developments among many researchers and academics as a "harbinger for emerging AI risk" (p. 1).

Richardson's Campaign Against Sex Robots (CASR), begun in 2015, helped to focus public attention on sex robot issues from a "feminist-humanist" perspective (Morris, 2018). CASR's stated goals include the following:

1. To abolish pornbots in the form of women and girls.
2. To offer an alternative, relational model of sex and sexuality informed by mutuality.
3. To challenge the normalisation of pornbots as substitutes for relationships with women.
4. To oppose the development of child sex-abuse dolls/robots as 'therapeutic' for paedophiles.
5. To offer up an alternative vision of technology where women and girls are centred and valued.
6. To work across the political spectrum with those who value the dignity of women and girls. (CASR, 2022)

The close linkages of humans (especially women) with sex robots provide some frightening prospects for the future and inspired the use of the term "pornbots" by the CASR rather than "sex robots." Objectification issues are difficult for societies, and having industries focus on how further to facilitate the development of individuated and custom-made sexual objects adds additional complexities. The complex gender dimensions of human–robot and human-AI interactions are indeed complex and are discussed in depth in Richardson's (2016, 2018) works along with a number of social robotics researchers (Döring et al., 2020; Escudero Perez, 2020; Ferrando, 2014). Winkle et al. (2021) outline the potential for development of "feminist robots" to deal with some of these gender-related challenges. Lupton (2015) expanded these discussions into the larger realm of sexually-themed health wearable devices. Some of these specific critiques focus on potential concerns involving sexual stereotypes and their perpetuation:

> … these technologies also serve to represent sexual activity and reproductive functions in certain defined and limited ways that work to perpetuate normative stereotypes and assumptions about women and men as sexual and reproductive subjects. Those apps that focus on sexual performance and competitiveness have the potential to incite anxiety and feelings of inadequacy in men, while women's bodies are further medicalised via the practices of intensive documentation and self-management these apps invite. (Lupton, 2015, pp. 448–449)

The roles of medical professionals in characterizing emerging concerns involved with sex robots are indeed growing, but do not yet have a strong presence in shaping public policy on these matters (Harper & Lievesley, 2020; McArthur & Twist, 2017). Despite potential health hazards, the configuration of these items as medical devices under the control of national-level health agencies is unlikely in the near future (Chatterjee, 2020). Analyses of the implications of these entities on health care have special complications, based on the lack of empirical evidence concerning harms and benefits:

> Evidence-based healthcare is at the core of medical professionalism and practice. The current dearth of information on health aspects of sexbots may relate to rapid commercial innovation, low sales, few direct consultations, failure to recognise and report health and social consequences for patients, or inadequate investment in research. (Cox-George & Bewley, 2018)

Negative aspects of human-technology interactions can be difficult to demonstrate in formal scientific ways to the extent at which they are broadly accepted in public policy circles; decades of research on the hypothesis that video games cause individuals to become violent have not produced the kinds of straightforward conclusions that would facilitate legislative and regulatory efforts. Comparable kinds of research on sex robots and related AI-enhanced sexual entities may also stall in part because of the rapid pace of development in the arena, with new forms of entities emerging before sold research can be conducted and peer reviewed.

COMMODIFICATION AND ADDICTION CONCERNS

As commodities, sex robots are engendering a range of problematic issues that are comparable to other kinds of sexual commodities, such as print and online pornography and sex toys (Troiano et al., 2020). Sex robot practices combine the potentials for expression of technical competence by the user with sexual competence; this combination can be perceived as empowering in some senses for the participant, just as the fusion of monetary clout with sexual expression can provide support for the proliferation of prostitution. In these early days of sex robot development, some forms of user technical know-how are often still required (if only to be able to customize the robot). Linkage of sex robots with human prostitution has brought unsettling insights about the current roles of sex workers

(Richardson, 2016), who often face considerable power and control issues as well as more directly sexual ones. The coalescing of these two powerful pursuits for users (achievements of technical and sexual performances), along with the acquisition and sharing of forms of technical know-how, has presented alluring prospects to many individuals.

The exterior appearance and surface operations of the robots play considerable roles in their functions, serving to create some sense of subjecthood; clothing, hair, and various surface characteristics can shape how they are perceived by their users (Friedman et al., 2021). These characteristics can easily be commodified, with various combinations costing certain amounts of money the way that consumers can pick out particular car colors for an extra charge. Moran (2019) contends that "Such robots' performance of subjecthood and identity therefore contribute to cultural instructions regarding how the racialised and gendered subject should be construed." The commodification of sexual activity that is engendered as part of sex robots is often coupled with commodification of various personality and identity indices, both of the users themselves and the chosen character (if any) of the sex robot. Simple sex robots (such as sex dolls) could indeed operate without extensive information about a human's response. However, new assortments of issues arise as various forms of human biometric indicators collected, analyzed, and utilized to produce robot feedback. As these individuated biometric indicators are collected by external parties (either the robot manufacturers and distributors, or healthcare providers), new issues concerning the match and fit of individuals with their sex robots are emerging, potentially fueled by insights from social robotics (such as Singh et al., 2021). Inappropriate or inadequate fit may lead to issues concerning the human user's wellbeing. Rich (2017) describes the related disorientation of individuals who do not conform to the expectations embodied in particular apps:

> What if one does not feel the way one is expected to, as portrayed in the normalized imagery accompanying the promotion of these apps (healthy, happy and in control)? What is the effect on one's subjectivity of one fails to achieve the ideals of these apps? What of those individuals who choose not to engage with these practices of self-surveillance and resist them? (p. 143)

Some sexual robotics and related AI-enhanced sex initiatives (such as chatbots) are part of technological approaches that are rooted in capitalizing on individuals' addictions (Eyal, 2014; Reith, 2018), where users cannot identify their limits and extricate themselves from unhealthy or dysfunctional use situations. Addiction has long been a factor in the societal issues involving the pornography industry, which fostered the excessive consumption on the part of a relatively small fraction of the population, with lesser consumption levels from other population sectors (Voss, 2015). Initiatives for making computer products and services "addictive" are a major concern in terms of the design and dissemination of computing technologies, but they are especially critical when it comes to sex robots, where certain kinds of overuse can have severe physical consequences. Identifying design efforts that are designed to foster addiction as opposed to those that are merely seeking to serve users can be difficult, but is required to ensure that the products involved do not negatively affect individuals' wellbeing. "Abuse" of sex robots is difficult to define: it can involve an amalgam of emotional, economic, and social concerns. Such issues as planned obsolescence (ending the support of particular versions of sex robots) can provide problems for user communities that are addicted to older versions.

Since sex robots are becoming an everyday commodity, some of the practical issues involving commodification need to be resolved, such as establishing warranties and determining when those warranties are voided (Banks, 2021). Actions that void a warranty can be characterized in terms of destruction, defacing, and manipulation; however, since some sexual activity may characterized in terms of such activity (if only on a symbolic level), legal authorities will have some complex decisions to make in this regard. Legal and regulatory approaches could have substantial impacts on the shape of sex robot issues, and considerable international differences are possible. As the kinds of injuries and possibly even deaths related to use of sex robots expand, and as companies that produce sex robots become richer, the attention of the legal and public policy communities to sex robot regulation initiatives may increase. In the legal realm, celebrity likeness issues are to be expected to loom large, as many individuals who choose to buy and use sex robots have them configured in ways that impinge upon the rights of individuals to have control over their likenesses.

The "New Pygmalion": Perspectives on Designer Sexuality

The themes of personal empowerment that are associated with human–robot-AI relationships can be complex and even contradictory. In contrast to the relatively-helpless positions of many individuals who abuse robots in workplaces (as described in previous chapters), sex robots often foster interactions that are designed on the surface to elevate individuals and enhance their power. However, despite these well-crafted suppositions, the information collected about the humans who are making the selections can be used in many ways for disempowerment and even exploitation. The creation of "artificial women and men" (following on the Pygmalion myth) has been a frequent theme in science fiction and other forms of mass entertainment, and has parallels in the creation of sex robots. The Pygmalion myth "tells the story of the revival of a lifeless sculpture by the male creator Pygmalion, who falls in love with her. This myth has been the subject of many works in literature and cinema" (Aksit & Favaro, 2019, p. 169). The Pygmalion myth (the centerpiece of such classic Western theatrical efforts as *My Fair Lady*) places the creator in a superior position over the object created, although it recognizes that this object can ultimately have significant impact on the creator's perspective and overall wellbeing. Since many robots have "creepy" influences on individuals, the Pygmalion-style impacts should be especially monitored; what makes sex robots appear dark and creepy to some individuals (such as their gathering of intimate personal information) may indeed be what can make them the most influential if not powerful in their overall implications. Few people have extensive experience in having sex with robots, and the way that these intense and formative experiences affect individuals (especially adolescents) is yet uncertain; fewer individuals have had the experience of creating and implementing a sex robot, so the "Pygmalion effects" have yet to be fully observed.

Designers are now just determining what kinds of choices people will want to make about the sex robots and other AI-related sexual entities (such as chatbots) in their lives. In much the way "cakemix" developers attempt to find out how much intervention and selection people desire in the recreational activity of baking a cake, sex robot developers are discovering what steps and decision sequences will entice individuals to construct their own ideal robotic mates. Consumers in the "new Pygmalion" approach are thus faced with a decision among various kinds of sex robots ("what features should their sex robots embody?") rather than basic decisions as to whether to spend time with a robot or a human being. Convincing individuals that a certain, perfectly configured and individualized sex robot would be the answer to their sexual woes is the kind of pursuit of which modern marketing methods are extraordinarily capable. The profitable "beauty industry" in which individuals are convinced that the right shade of lipstick will make them more attractive and the psychotropic drug industry that convinces individuals that certain combinations of medications will make them happy are both examples of how questions have been framed for consumers in ways that lock them into ultimately-ineffective decision patterns (De Regt et al., 2020; Gardner, 2003).

The theme of robots apparently "outclassing" humans in problematic ways (utilized throughout this book) has special implications for sexuality, especially in situations in which individuals can design and fine-tune specific features of the sexual robots with whom one is engaging. The robotic companions involved are often framed in personal narratives as well as advertisements as "outclassing" human companions by being part of the clean and intellectually superior high-tech sphere (rather than the messy and imprecise human realm). Whether or not individuals seek to use sex robots as ways to establish comparisons with humans in their lives, the fact is that many of the themes and images they will integrate into their robotic activity stem from the human-to-human encounters they experienced throughout their lives. Some of the intimate competitions that robots and AI-enhanced entities are involved in present implied contests between humans and robots in their capacities to fulfill certain stereotypical sexual and social functions; these functions may change in character as people experiment with sex robots and exploit their abilities to do things repetitiously without apparent signs of boredom. The young people who are rejected in a relationship with another individual because they do not perform as well as sex robots in certain erotic activities have every reason to be confused. The fear of humans "becoming a robot" as a function of competitions with sex robots has been expressed in an assortment of venues as human individuals endeavor to find a place in a social order that includes AI and robots (Clinton et al., 2018). Sexual activity with robots may indeed "be morally *disruptive* and cause moral revolutions" (Danaher, 2020), and "loving robots [may] change love" (Sætra, 2021, p. 1), though in some disquieting and often disturbing ways.

Blurring the boundaries between people and robots can present challenges even when direct sexual contact is not involved. Williams extols the delights of being with sex robots in a way that emphasizes companionship and "shared pleasures":

> Probably one of the biggest beneficiaries to emerge from the advancement of AI tied to robotics could be the sex and companion robot directed at both men and women worldwide. The results are hyper-realistic models that breathe, have orgasms and a heartbeat. Thanks to advanced AI, these sex robots can hold a conversation about movies, music or books. They can tell jokes, remember your brother's name and, of course, will have sex with you whenever you want. (Williams, 2021, para. 19)

The construction of such "friendship" between humans and robots can have an impact on the very concept of companionship (Elder, 2017), affecting the social standing of humans and fostering the "robotization" of people.

Just as sexting expanded the range of sexual expression through technological means (Oravec, 2012), sex robots are presenting new potentials for human sexuality. From a dramaturgical perspective, sex robots can change the audience and both the backstage and frontstage venues for sexuality, and in some ways alter how the human participant is viewed in a reverse-Pygmalion manner. However, the users of the sex robot construe and interpret their interactions with the robots, the potential for backstage quantitative manipulation by the developers of the robots looms large; developers could be characterizing and shaping the sexual interactions quite differently than they would be in a straightforward clinical context, emphasizing certain kinds of statistics (such as pseudo comparisons with other individuals) and even introducing strategic errors and omissions. The human participants who believe that they are in control of the sex robot development and use situation are themselves being shaped and controlled. In many applications, sex robot usage may increase the kinds and levels of quantification applied to sexuality, just as the current array of "quantified self" health wearables have shaped some forms of physical exercise and even sleep (Lupton, 2015; Oravec, 2020, 2022). The sex robots' creators and users may thus be shaped by their creations.

This chapter indeed cannot take on the full range of unsettling prospects for sex robot applications. The full spectrum of psychological impacts of sex robot usage includes high-tech harassment in robotic, AI-enhanced, and metaverse spaces (Basu, 2021); for example, Taylor (2022) reports in a business news magazine how "men are creating AI girlfriends, verbally abusing them, and bragging about it on Reddit." In another problematic context, fidelity issues in robot-human sexual situations are discussed in Rothstein et al. (2021). The consent-related matters such as the appropriation of others' images and characterizations for sexual purposes are already attracting considerable ethical and legal attention (Ley & Rambukkana, 2021). The ease with which sexuality is being conveyed to children via widely-available sex robot games and toys is also of concern, including in some popular game platforms (Clayton & Dyer, 2022). Sex robot development and proliferation are serving to define and commercialize socially acceptable sexual behavior in a kind of "sexual engineering," as well as expand the scope of personal privacy and image appropriation concerns in intimate realms.

Some Conclusions and Reflections: The Dark Sides of Human–Robot Communion

The choice of sex robots to illustrate some dark side and creepiness issues for this book was reinforced by a growing literature of popular culture and research narratives that deal with their negative concerns (Cavanagh, 2021; Liberati, 2021; Richardson, 2016). However positively or negatively they are characterized, sex robots and other AI-related sexually-themed entities are playing roles in the normalization of robotics in everyday life, the way that we are almost numbed to the prospects of further robot and AI incursion. In some contexts, these entities are fostering the perceived elevation of robots to higher social and moral levels as humans commune with and even marry them; for some individuals, robots and AI are themselves characterized as the superior entities. The prospects for proliferation of sex robots and related technosex objects underscore the emerging controversies in whether or how to segregate and contain the negative applications of technology, or the "dark side," from approaches that are apparently more functional and appropriate. Sexual activity with robots may indeed provide positive outlets and be an aspect of healthy personal expression in certain contexts. However, sexual activity is also linked with lengthy legacies of bodily dissociation, repression, sadism, masochism, pedophilia, misogyny, transphobia, and a wide range of other cultural threads that can be employed opportunistically in the development and marketing of sex robots, as well as in how their intended uses are expanded by their consumers. For example, rough and abusive sex with robots could soon, in many circumstances, be a way of ventilating anxieties toward the transportation, manufacturing, and service robots that are unfortunately controlling individuals' lives. Sex robots and other AI-enhanced sexual technologies may indeed enable people to discover new personal identities and social roles; however, warnings are needed as those exploratory efforts are abused by corporations, malicious individuals, and government agencies. Individuals' decision making in conjunction with sexual activities often tends not to be coherent, and this vulnerable period of time in human life can be ripe for exploitation, either for the inappropriate elicitation of personal information, for unhealthy lifestyle selections, or for unsuitable economic choices.

Along with the glowing support of sex robots from some advocates, serious and sustained opposition to sex robot and related technosex developments has emerged in the past decades. A clinical perspective toward sex robots often presents the use of these entities as part of a healthy sexual release, one that can be precisely measured and analyzed with various sensors and biometric devices. However, sexual contact and ideation can have a wide spectrum of influences on individuals' wellbeing (as well as cultural trends), including profoundly negative ones. One of the varieties of "creepiness" associated with sex robots is that the romantic and personal sides of some sex robot contact are being externally controlled and narrated, largely by the corporate interests that designed and distributed the robot. As discussed throughout this book, the complexities of robots enable the production of a wide assortment of "dark patterns" by developers that can deviate from what is openly presented to the user. Whether these dark pattern manipulations will lead the sex robot users to buy more corporate-sponsored items or to manifest some political or social expressions depends on the robot and the situation. Many of the dark patterns are so subtle that courts and legislators will not be able to identify and defend against them, just as so much flirtatious behavior has inappropriate and even evil intent though it is well within the bounds of what society presumably deems acceptable.

The widespread proliferation and use of sex robot and AI-enhanced sex technologies can potentially affect how individuals characterize robotics and AI themselves as well as potentially alter their perspectives toward each other as human beings. A "slippery slope" of normalization of sex robot behaviors can serve to make humans less vigilant and aware of real hazards on the job with manufacturing or service robots or on the roadway with autonomous vehicles. Design perspectives for robotics in the future could well change as individuals' comfort with sex robots alters many users' assumptions and characterizations of how robots operate. The marketing and promotion of sex robots and related products are likely to continue to include exaggerated expression of how advanced are the AI-related capabilities of the products, leading to an expansion in AI-related hype (as described in a previous chapter). Research could help to identify the ways that the sex robots appear creepy, in part to catch emerging "danger signals" that the technologies have negative influences on individuals (Woźniak et al., 2021). However, the more generally accepted applications of sex robot technology are often difficult to distinguish from the "creepier" and potentially illegal (in terms of informed consent and intellectual property) ones.

Given the social and moral complexities involving sex robot development, one can expect that increased levels of concern will be expressed by various professional organizations and special interest groups. Many detailed and often heroic initiatives of researchers in high-tech fields are being placed toward sex robot development (Giger et al., 2019), and this societal investment in technosex products and services could well be directed toward other kinds of social goals and unmet needs. The possible courses of action of individual researchers who do not want these kinds of technological applications to proliferate can be limited, constrained by the fact that sex robots are not yet being taken seriously in many political and technical realms. In contrast, the defense of sex robots by those who have some emotional or addictive tie to them is also likely to be a growing issue, since the levels of personal identification with and advocacy for the machines are likely to be substantial. The Pygmalion-style bonds that individuals have with their creations can be considerable, as described through the centuries in theatrical and historical narratives. The perceived inferiority of one's human sexual partner may indeed propel many individuals into interacting with a perceptibly "cleaner," more intelligent, and more compliant entity the characteristics of which have been specially tailored to one's own profile (however honest the numbers may be) and expressed preferences.

Sex robot discourse provides clues as to how other kinds of emerging robot and AI applications will be construed, such as military and security robots (with their control-related themes). We should all attempt to explore our own symbolic and aesthetic relationships with robots in order to understand some of these interactions at a personal level. Since sex robots have become available on the market, a great deal of attention has been directed to their prospects in popular and academic discourse. However fanciful and off-the-mark at times, science fiction may indeed also assist us in imagining the directions our intimate futures may take (as in the recent efforts of Campbell [2021] and Wilson [2021] as well as Asimov [1983]). These discourses add to the legacy that is associated with robots and other AI-related entities, so sex robot topics are of relevance to a wide variety of robotic and other high-tech initiatives. Individuals who are compliant enough to buy the newest sex robot and have aspects of their intimate lives recorded and shaped with AI may be malleable enough to work for organizations that are designed for the efficient operation of robot hardware and live in regions that are controlled by military and police robots.

References

Aksit, O. O., & Favaro, A. (2019, December). Pygmalion myth and artificial women in contemporary science fiction films. In *Proceedings of Seventh International Mediterranean Social Sciences Congress* (MECAS VII) (pp. 169–176).

Asimov, I. (1983). *The robots of dawn.* Doubleday.

Banks, J. (2021). From warranty voids to uprising advocacy: Human action and the perceived moral patiency of social robots. *Frontiers in Robotics and AI, 8.* https://doi.org/10.3389/frobt.2021.670503

Basu, T. (2021, December 16). The metaverse has a groping problem already. A woman was sexually harassed on Meta's VR social media platform. *Technology Review.* https://www.technologyreview.com/2021/12/16/1042516/the-metaverse-has-a-groping-problem/

Bendel, O. (2015). Surgical, therapeutic, nursing and sex robots in machine and information ethics. In *Machine medical ethics* (pp. 17–32). Springer.

Björkas, R., & Larsson, M. (2021). Sex dolls in the Swedish media discourse: Intimacy, sexuality, and technology. *Sexuality & Culture, 25,* 1227–1248. https://doi.org/10.1007/s12119-021-09829-6

Brooks, R. (2021a). Tomorrow's moral panic will be just like yesterday's. In *Artificial Intimacy* (pp. 152–176). Columbia University Press.

Brooks, R. (2021b, May 8). Click and connect: The rise of 'artificial intimacies.' *The Australian.* https://www.theaustralian.com.au/weekend-australian-magazine/click-and-connect-the-rise-of-artificial-intimacies/news-story/a8602a9b626b9d3e154c913646fef3d1

Brown, R., & Shelling, J. (2019). Exploring the implications of child sex dolls. *Trends and Issues in Crime and Criminal Justice [Electronic Resource], 570*, 1–13.

CASR (Campaign Against Sex Robots). (2022). https://campaignagainstsexrobots.org/

Campbell, A. (2021). *The love makers*. Goldsmiths.

Card, K. (2021, December 16). Westworld-style sex robot factory promises 'exciting' new dolls in 2022. *Daily Star.* https://www.dailystar.co.uk/news/weird-news/westworld-style-sex-robot-factory-25711385

Cavanagh, N. (2021, March 21). Sex robots could soon be "cloned" to look like your dead partner thanks to groundbreaking 3-D scans. *The Sun,* 26 March. https://www.thesun.co.uk/tech/14469427/sex-robots-cloned-dead-partner-3d/

Chatterjee, B. B. (2020). Child sex dolls and robots: Challenging the boundaries of the child protection framework. *International Review of Law, Computers & Technology, 34*(1), 22–43.

Cheok, A. D., Karunanayaka, K., & Zhang, E. Y. (2017). Human-robot love and sex relationships. In P. Lin, K. Abney, R. Jenkins (Eds.), *Robot ethics: From autonomous cars to artificial intelligence* (pp. 193–220).

Cheok, A. D., & Zhang, E. Y. (2019). Sex and a history of sex technologies. In *Human–robot intimate relationships* (pp. 23–32). Springer.

Clayton, J., & Dyer, J. (2022, February 15). Roblox: The children's game with a sex problem. *BBC News.* https://www.bbc.com/news/technology-60314572

Clinton, M., Madi, M., Doumit, M., Ezzeddine, S., & Rizk, U. (2018). "My greatest fear is becoming a robot": The paradox of transitioning to nursing practice in Lebanon. *SAGE Open, 8*(2), 2158244018782565.

Coeckelbergh, M. (2010). Artificial companions: Empathy and vulnerability mirroring in human-robot relations. *Studies in Ethics, Law, and Technology, 4*(3), 1–17. https://doi.org/10.2202/1941-6008.1126

Coopersmith, J. (2006). Does your mother know what you really do? The changing nature and image of computer-based pornography. *History and Technology, 22*(1), 1–25.

Cotton, B. (2018). Sex Robots—A disturbing look inside one of the world's fastest growing industries. *Business Leader.* https://www.businessleader.co.uk/sex-robots-a-disturbing-look-inside-one-of-the-worlds-fastest-growing-industries/53749/

Cox-George, C., & Bewley, S. (2018). I, sex robot: The health implications of the sex robot industry. *BMJ: Sexual and Reproductive Health, 44*(3). https://srh.bmj.com/content/44/3/161.short

Danaher, J. 2020. Robots and moral revolutions. In *Culturally sustainable social robotics*. IOS Press.

De Regt, A., Montecchi, M., & Ferguson, S. L. (2020). A false image of health: How fake news and pseudo-facts spread in the health and beauty industry. *Journal of Product & Brand Management, 29*(2), 168–179. https://doi.org/10.1108/JPBM-12-2018-2180

Di Nucci, E. (2016). Sexual rights, disability and sex robots. In J. Danaher & N. McArthur (Eds.), *Sex robots* (pp. 73–88). MIT Press

DiTecco, D. (2020). *New technology, same old stigma: An analysis of feminist discourses and sex work stigma in sex robot media* (Doctoral dissertation, Carleton University). https://doi.org/10.22215/etd/2020-14051

Döring, N., Mohseni, M. R., & Walter, R. (2020). Design, use, and effects of sex dolls and sex robots: Scoping review. *Journal of Medical Internet Research, 22*(7), e18551.

Döring, N., & Pöschl, S. (2018). Sex toys, sex dolls, sex robots: Our under-researched bed-fellows. *Sexologies, 27*(3), e51–e55.

Dubé, S., & Anctil, D. (2020). Foundations of erobotics. *International Journal of Social Robotics, 13*, 1205–1233. https://doi.org/10.1007/s12369-020-00706-0

Dundes, A. (2017). *Cracking jokes: Studies of sick humor cycles & stereotypes*. Quid Pro Books.

Eden, A. H., Moor, J. H., Søraker, J. H., & Steinhart, E. (2012). *Singularity hypotheses*. The Frontiers Collection. Springer.

Effoduh, J. O. (2021). The legitimization of customized sex robots in the age of COVID-19. *Intellectual Property Journal, 33*(2), 161–181.

Elder, A. M. (2017). *Friendship, robots, and social media: False friends and second selves*. Routledge.

Escudero Perez, J. (2020). "An AI doesn't need a gender" (but it's still assigned one): Paradigm shift of the artificially created woman in film. *Feminist Media Studies, 20*(3), 325–340.

Eyal, N. (2014). *Hooked: How to build habit-forming products*. Penguin.

Ferrando, F. (2014). Is the post-human a post-woman? Cyborgs, robots, artificial intelligence and the futures of gender: A case study. *European Journal of Futures Research, 2*(1), 43–60.

Flore, J., & Pienaar, K. (2020). Data-driven intimacy: Emerging technologies in the (re)making of sexual subjects and 'healthy' sexuality. *Health Sociology Review, 29*(3), 279–293.

Friedman, N., Love, K., LC, R., Sabin, J. E., Hoffman, G., & Ju, W. (2021, June). What robots need from clothing. In *Designing Interactive Systems Conference 2021* (pp. 1345–1355).

Galaitsi, S. E., Hendren, C., Trump, B., & Linkov, I. (2019). Sex robots—A harbinger for emerging AI risk. *Frontiers of Artificial Intelligence*. https://doi.org/10.3389/frai.2019.00027

Gardner, P. (2003). Distorted packaging: Marketing depression as illness, drugs as cure. *Journal of Medical Humanities, 24*(1), 105–130.

Gersen, J. S. (2019). Sex Lex machina. *Columbia Law Review, 119*(7), 1793–1810.

Giger, J. C., Piçarra, N., Alves-Oliveira, P., Oliveira, R., & Arriaga, P. (2019). Humanization of robots: Is it really such a good idea? *Human Behavior and Emerging Technologies, 1*(2), 111–123.

Goffman, E. (1959). *The presentation of self in everyday life.* Doubleday.

Guarino, B. (2018, June 4). New report finds no evidence that having sex with robots is healthy. *The Washington Post.* https://www.washingtonpost.com/news/speaking-of-science/wp/2018/06/04/theres-no-evidence-that-having-sex-with-robots-is-healthy-new-report-finds/

Harper, C. A., & Lievesley, R. (2020). Sex doll ownership: An agenda for research. *Current Psychiatry Reports, 22*(10), 1–8.

Hasse, C. (2019). The Vitruvian robot. *AI & Society, 34*(1), 91–93.

Ide, W. (2021, August 14). I'm Your Man review—Domestic bliss with a humanoid. *The Guardian.* https://www.theguardian.com/film/2021/aug/14/im-your-man-review-dan-stevens-robot-berlin

Inniss, L. K. B. (2001). Bicentennial man—The new millennium assimilationism and the foreigner among us. *Rutgers Law Review, 54*, 1101–1132.

Jecker, N. S. (2021). Nothing to be ashamed of: Sex robots for older adults with disabilities. *Journal of Medical Ethics, 47*(1), 26–32.

Kakoudaki, D. (2014). *Anatomy of a robot.* Rutgers University Press.

Kermode, M. (2021, December 26). *Titane* review—Agathe Rousselle is extraordinary in Palme d'Or-winning body horror. *The Guardian.* https://www.theguardian.com/film/2021/dec/26/titane-review-julia-ducournau-palme-dor-cronenberg-agathe-rousselle

Klein, W. E., & Lin, V. W. (2018). "Sex robots" revisited: A reply to the campaign against sex robots. *ACM SIGCAS Computers and Society, 47*(4), 107–121.

Kubes, T. (2019). Bypassing the uncanny valley: Sex robots and robot sex beyond mimicry. In *Feminist philosophy of technology* (pp. 59–73). JB Metzler.

Lacey, C., & Caudwell, C. (2019, March). Cuteness as a 'Dark Pattern' in home robots. In *2019 14th ACM/IEEE International Conference on Human-Robot Interaction (HRI)* (pp. 374–381). IEEE.

Lee, J. (2017). *Sex robots: The future of desire.* Springer.

Levy, D. (2007). *Love and sex with robots: The evolution of human-robot relationships.* Harper.

Ley, M., & Rambukkana, N. (2021). Touching at a distance: Digital intimacies, haptic platforms, and the ethics of consent. *Science and Engineering Ethics, 27*(5), 1–17.

Liberati, N. (2021). Phenomenology and sex robots: A phenomenological analysis of sex Robots, threesomes, and love relationships. *International Journal of Technoethics (IJT), 12*(2), 86–97.

Liu, J. (2021). Social robots as the bride? Understanding the construction of gender in a Japanese social robot product. *Human-Machine Communication, 2*(1), 105–120.

Luguri, J., & Strahilevitz, L. J. (2021). Shining a light on dark patterns. *Journal of Legal Analysis, 13*(1), 43–109.

Lunceford, B. (2018). Love, emotion and the singularity. *Information, 9*(9), 221–231.

Lupton, D. (2015). Quantified sex: A critical analysis of sexual and reproductive self-tracking using apps. *Culture, Health & Sexuality, 17*(4), 440–453.

Mainenti, D. C. (2020). Sex robot technology and the Narrative Policy Framework (NPF): A relationship in the making? *Paladyn, Journal of Behavioral Robotics, 11*(1), 390–403. https://doi.org/10.1515/pjbr-2020-0022

Maldonato, N. M., Valerio, P., Bottone, M., Sperandeo, R., Scandurra, C., Punzo, C., Muzii, B., D'Andrea, S., & Eposito, A. (2021). The desiring algorithm. The sex appeal of the inorganic. In *Progresses in Artificial Intelligence and Neural Systems* (pp. 607–613). Springer, Singapore.

Maras, M. H., & Shapiro, L. R. (2017). Child sex dolls and robots: More than just an uncanny valley. *Journal of Internet Law, 21*(5), 3–21.

Martin, F. (2018, October 17). Is this the end for a sex robot brothel in Houston? *Houston Public Media.* https://www.houstonpublicmedia.org/articles/news/in-depth/2018/10/17/308292/is-this-the-end-for-a-sex-robot-brothel-in-houston/

McArthur, N., & Twist, M. L. (2017). The rise of digisexuality: Therapeutic challenges and possibilities. *Sexual and Relationship Therapy, 32*(3–4), 334–344.

McBride, J. (2021). Climate change, global population growth, and humanoid robots. *Journal of Future Robot Life, 2* (Preprint), 23–41.

Moran, J. C. (2019). Programming power and the power of programming: An analysis of racialised and gendered sex robots. In *Feminist philosophy of technology* (pp. 39–57). JB Metzler.

Morris, A. (2018, September 26). Meet the activist fighting sex robots. *Forbes.* https://www.forbes.com/sites/andreamorris/2018/09/26/meet-the-activist-fighting-sex-robots/?sh=32307c4f6e79

Nagy, J. (2021). Pink chat: Networked sex work before the internet. *Technology and Culture, 62*(1), 57–81.

Nature of Things. (2021, November 18). 'I love her and see her as a real woman.' Meet a man who 'married' an artificial intelligence hologram.

CBC.Ca. https://www.cbc.ca/documentaries/the-nature-of-things/i-love-her-and-see-her-as-a-real-woman-meet-a-man-who-married-an-artificial-intelligence-hologram-1.6253767

Nazaroff, D. (2021, April 30). Virtual sex and dollbots could lead to exploitation, academic warns. *UNSW.* https://newsroom.unsw.edu.au/news/science-tech/virtual-sex-and-dollbots-could-lead-exploitation-academic-warns

Nevett, J. (2018, January 24). UK's first sex doll brothel faces backlash as prostitutes fear 'going out of business'. *Daily Star* (Online). https://www.dailystar.co.uk/news/latest-news/676702/uk-sex-doll-brothel-lovedoll-prostitutes-backlash-business-gateshead

Nield, D. (2021, August 21). Legal researchers weigh in on how future laws should deal with the rise of the sexbots. *ScienceAlert.* https://www.sciencealert.com/legal-researchers-weigh-in-on-how-future-laws-should-deal-with-the-rise-of-the-sexbots

O'Brien, S. (2021, August 21). Sex workers helped popularize OnlyFans. Now their future on the platform is uncertain. *CNN.com.* https://www.cnn.com/2021/08/21/tech/onlyfans-creators-react-sexually-explicit-ban/index.html

Oravec, J. A. (2012). The ethics of sexting: Issues involving consent and the production of intimate content. In D. Heider & A. Massanari (Eds.), *Digital ethics: Research and practice* (pp. 129–145). Peter Lang.

Oravec, J. A. (2014). Craigslist in crisis: Issues of censorship and moral panic in the context of online communities. *International Journal of the Academic Business World, 8*(2), 1–11.

Oravec, J. A. (2020). Digital iatrogenesis and workplace marginalization: Some ethical issues involving self-tracking medical technologies. *Information, Communication & Society, 23*(14), 2030–2046.

Oravec, J. A. (2022). The emergence of "truth machines"?: Artificial intelligence approaches to lie detection. *Ethics and Information Technology, 24*(1), 1–10.

Paasonen, S. (2011). *Carnal resonance: Affect and online pornography.* MIT Press.

Patient, D. (2019. November 7). Swiss brothel swaps women for sex robots after demand soars for £70 romps. *Daily Star.* https://www.dailystar.co.uk/news/world-news/swiss-brothel-swaps-women-sex-20836948

Peeters, A., & Haselager, P. (2021). Designing virtuous sex robots. *International Journal of Social Robotics, 13*(1), 55–66.

'Proud robosexual' plans to marry robot when it's legal. (2016, December 25). www.inquisitr.com. https://www.inquisitr.com/3826210/proud-robosexual-plans-to-marry-robot-when-its-legal/

Ray, A. (2018). *Compassionate artificial intelligence: Frameworks and algorithms.* Compassionate AI Lab (An Imprint of Inner Light Publishers).

Reith, G. (2018). *Addictive consumption: Capitalism, modernity and excess.* Routledge.

Rich, E. (2017). Childhood, surveillance and mHealth technologies. In E. Taylor & T. Rooney (Eds.), *Surveillance futures: Social and ethical implications of new technologies for children and young people* (pp. 132–145). Routledge.

Richardson, K. (2015). The asymmetrical 'relationship' parallels between prostitution and the development of sex robots. *ACM SIGCAS Computers and Society, 45*(3), 290–293.

Richardson, K. (2016). Sex robot matters: Slavery, the prostituted, and the rights of machines. *IEEE Technology and Society Magazine, 35*(2), 46–53.

Richardson, K. (2018). *Sex robots: The end of love*. Polity Press.

Rothstein, N. J., Connolly, D. H., de Visser, E. J., & Phillips, E. (2021, March). Perceptions of infidelity with sex robots. In *Proceedings of the 2021 ACM/IEEE International Conference on Human-Robot Interaction* (pp. 129–139).

Sætra, H. S. (2021). Loving robots changing love: Towards a practical deficiency-love. *Journal of Future Robot Life* (Preprint), 1–19.

Salas-Rodriguez, I. (2021, December 2). First lesbian sex robots may be coming soon as sexbots start being programmed to be gay. *The Sun*. https://www.the-sun.com/news/4193896/first-lesbian-sex-robots-coming-soon/

Singh, K. J., Kapoor, D. S., & Sohi, B. S. (2021). Selecting social robot by understanding human–robot interaction. In *International Conference on Innovative Computing and Communications* (pp. 203–213). Springer, Singapore.

Steele, B. J. (2021). "A catharsis for anxieties": Insights from Goffman on the politics of humour. *Global Society, 35*(1), 102–116.

Székely, M., Powell, H., Vannucci, F., Rea, F., Sciutti, A., & Michael, J. (2019). The perception of a robot partner's effort elicits a sense of commitment to human-robot interaction. *Interaction Studies, 20*(2), 234–255.

Taylor, A. (2022, January, 19). Men are creating AI girlfriends, verbally abusing them, and bragging about it on Reddit. *Fortune*. https://fortune.com/2022/01/19/chatbots-ai-girlfriends-verbal-abuse-reddit/

Torre, B. (2021, May 29). Woman cloned into sex doll so realistic you'd need microscope to tell them apart. *Daily Star*. https://www.dailystar.co.uk/news/weird-news/woman-cloned-sex-doll-realistic-24212090

Tosin. (2017, April 28). N6 million sex robot can quote Shakespeare, remember birthdays. *The Herald* (Nigeria). https://www.herald.ng/n6-million-sex-robot-can-quote-shakespeare-remember-birthdays/

Troiano, G. M., Wood, M., & Harteveld, C. (2020, April). And this, kids, is how I met your mother: Consumerist, mundane, and uncanny futures with sex robots. In *Proceedings of the 2020 CHI Conference on Human Factors in Computing Systems* (pp. 1–17). ACM.

van Grunsven, J., & van Wynsberghe, A. (2019). A semblance of aliveness: How the peculiar embodiment of sex robots will matter. *Techne: Research in Philosophy and Technology, 23*(3), 290–317.

Voss, G. (2015). *Stigma and the shaping of the pornography industry*. Routledge.

Wade-Palmer, C. (2021). Desperate sex robot fans pledge to give their own money to speed up AI doll advancements. *Daily Star*. https://www.dailystar.co.uk/news/weird-news/desperate-sex-robot-fans-pledge-24372710

Watt, M. C., Maitland, R. A., & Gallagher, C. E. (2017). A case of the "heeby jeebies": An examination of intuitive judgements of "creepiness." *Canadian Journal of Behavioural Science/Revue canadienne des sciences du comportement, 49*(1), 58.

Williams, M. F. (2021, July 19). Sex tech and the rise of AI robotics in the bedroom: 9 investment ideas. *Financial News Now*. https://financial-news-now.com/sex-tech-and-the-rise-of-ai-robotics-in-the-bedroom-9-investment-ideas/

Wilson, D. (2021). *Ride, sally, ride*. Canon Press.

Winkle, K., Melsión, G. I., McMillan, D., & Leite, I. (2021, March). Boosting robot credibility and challenging gender norms in responding to abusive behaviour: A case for feminist robots. In *Companion of the 2021 ACM/IEEE International Conference on Human-Robot Interaction* (pp. 29–37).

Wolf, J. M. (2012). *Technosexuality: Technology, sexuality, and convergence* (Doctoral dissertation, Syracuse University).

Woźniak, P. W., Karolus, J., Lang, F., Eckerth, C., Schöning, J., Rogers, Y., & Niess, J. (2021, May). Creepy technology: What is it and how do you measure it? In *Proceedings of the 2021 CHI Conference on Human Factors in Computing Systems* (pp. 1–13).

Yanke, G. (2021). Tying the knot with a robot: Legal and philosophical foundations for human–artificial intelligence matrimony. *AI & Society, 36*(2), 417–427.

Zara, G., Veggi, S., & Farrington, D. P. (2021). Sexbots as synthetic companions: Comparing attitudes of official sex offenders and non-offenders. *International Journal of Social Robotics, 14*, 479–498.

CHAPTER 5

The Long Robotic Arm of the Law: Emerging Police, Military, Militia, Security, and Other Compulsory Robots

Many kinds of robots, autonomous vehicles, and other AI-enhanced entities have been developed to fill military, police, and security missions, with capabilities including facial recognition, thermal imaging, and various other biometric identifications (Kelly, 2018; Rezende, 2020). Increasingly, they are designed to incapacitate or kill humans in certain circumstances without the direct and immediate control of a human being. This chapter focuses on the opportunistic uses of these entities to control populations, as well as the growing resistance to such initiatives by many human rights activists, academics, international organizations, and others concerned about the future prospects for humanity. Ploumis (2022) relates how "these systems are already being used and have the capability to successfully deal with threats faster than humans" (p. 1) providing little time for policy makers and civilians to attempt to intervene in their development, proliferation, and usage. One reason for caution is how the existence of these systems can increase public anxiety. The public's enhanced senses of fear and dread may be hard to quantify, but can emerge in various demonstrated mental health conditions and behavioral patterns (Sharkey, 2019), with children being affected as well as adults (Wittes & Blum, 2015).

Robotic and other AI-enhanced systems are promoted as reducing the potential for direct attacks on human soldiers and security workers. The ideal of designing our military systems "so military action will be clean—our robots versus their robots, and may the best cybernetic force win" (Reynolds, 1991, p. 723) has been discussed for decades, presenting the possibility of armed conflict conducted without human soldiers as well as in remote regions of the Earth and even outer space (as with the US Space Force efforts). The stated benefits of robotic and AI approaches to military and police work have included the potential to remove personnel from venues in which they can be physically harmed. For example, Osborn (2021) relates how the US Army "is ramping up investments in robotic systems that could keep troops out of harm's way and serve as force multipliers on future battlefields." Other AI-related interventions are also changing military operations: various AI-enhanced lie detection systems are promoted as being able to determine whether enemy soldiers are lying, and can often be administered at a distance (Oravec, 2022). Opponents of integrating autonomous systems into warfare argue that the costs of having unreliable systems that are equipped to kill humans outweigh the supposed benefits in the protections for military and security personnel. Such potentials as intelligent military and security systems should also inspire advanced thinking about how to eliminate armed conflict altogether, rather than design new kinds of warfare (albeit "clean" warfare).

The construct of the "killer robot" has an extensive legacy in international military negotiations and fictional narratives as well as in newspaper and social media accounts, but these entities are only one of an assortment of AI-enhanced systems used in military and security contexts. The acronym "LAWS" (Lethal Autonomous Weapon Systems) has served to moderate the power of the killer robot label in some discourse on the topic, characterizing "systems capable of identifying targets and using deadly force without human control… also referred to as autonomous weapons systems, autonomous military systems, lethal autonomous robotics, and 'killer robots'" (Kelly, 2018). The lethality of the technologies involved has attracted the attention of many human rights advocates and international agencies, as well as religious leaders such as the Pope (Rosert & Sauer, 2021; Sadowski, 2021). The term "slaughterbots" has also been injected into these discussions, in part because of a film produced by The Future of Life Institute that presents an assortment of futuristic killer robot scenarios (Weber, 2021). Military forces in various nations have already utilized non-combat robots in a variety of ways, for example to "scout enemy fire from around the corner; scan buildings for spot threats; carry teams' ammo, water, gear, batteries; use thermal camera and chemical sensor to report back on city sewer system; and even scour for explosives or enemy fighters in the dark" (Brown-Gaston & Arora, 2021, p. 1). The autonomous killer robot goes beyond support capabilities to a level of operations in which it takes on initiatives, presumably within certain bounds, to engage directly in combat.

The perceived inevitability of robots and other AI-enhanced entities in instituting autonomous control in military and security settings presents stunning scenarios for the future, ones in which humans become less equipped to make choices about their own safety and future prospects. The complexity of the operations of some military robots could make their oversight by humans exceedingly difficult, even for those well-versed in their operations. A former design engineer for these systems related how accidents can occur in the midst of such complexity:

> The likelihood of a disaster is in proportion to how many of these machines will be in a particular area at once. What you are looking at are possible atrocities and unlawful killings even under laws of warfare, especially if hundreds or thousands of these machines are deployed... There could be large-scale accidents because these things will start to behave in unexpected ways. Which is why any advanced weapons systems should be subject to meaningful human control, otherwise they have to be banned because they are far too unpredictable and dangerous. (McDonald, 2019)

The roles played by military and police robots can be comparable to those of human military and police agents, but with the latter the potential for some kind of commonsense or humane reasoning is still a factor in particular situations (despite the apparent brutality associated with some humans' conduct).

Accounts of how these robots and other entities will actually work in practice are still emerging. News narratives such as the following about China's reported use of killer robots are tentative in tone, but emphasize the potential autonomous operations of the technologies:

> China has deployed unmanned ground vehicles (UGV) to the [Indian border] region in an effort to bolster its position along the border. According to the report, China's People's Liberation Army acquired the UGVs in 2014 but has largely neglected their use until now. Presently, the Chinese military utilizes two different types of UGVs, the Sharp Claw

and the Mule-200. The Sharp Claw can be operated remotely in addition to performing autonomous behaviour, according to *National Interest*. The Mule-200 can perform unmanned deliveries or operate firearms, including mounted guns. India's *WION News* reports that at least 400 of the robotic weapons platforms have been deployed along the border. (Cheong, 2022, para. 2)

According to McDonald (2019), some of the US and Russian autonomous weapon systems include the following:

- The US Navy's AN-2 Anaconda gunboat, which is being developed as a "completely autonomous watercraft equipped with artificial intelligence capabilities" and can "loiter in an area for long periods of time without human intervention".
- Russia's T-14 Armata tank, which is being worked on to make it completely unmanned and autonomous. It is being designed to respond to incoming fire independent of any tank crew inside.
- The US Pentagon has hailed the Sea Hunter autonomous warship as a major advance in robotic warfare. An unarmed 40-metre-long prototype has been launched that can cruise the ocean's surface without any crew for two to three months at a time.

The role of AI in these autonomous systems makes it difficult for humans to be meaningfully involved in their operations. Their complexities make effective human intervention in adjusting for special circumstances or in fixing corrupted processes less of a possibility. Osborn's (2021) account underscores some of these complexities: "Incoming data can be gathered, aggregated, and then bounced off of an AI-empowered database to quickly make comparisons, assess new information in relation to established information or past contingencies, perform analysis and make recommendations to human commanders regarding what an optimal approach might be. This enabled speed, and the near-simultaneous organization and categorization of multiple variables in relation to one another" (para. 4).

Combination of autonomous capabilities and a remote-controlled option is provided in some military systems, as in the Jaeger-C robots which were purchased by the Australian Army in 2021:

The wheeled Jaeger-C is a small machine with a low profile designed to attack from ambush… the makers note it can be remote-controlled or operate "autonomously with image analysis and trained models linked to robotic actions," according to a report in *Overt Defense*. This sounds very much like the sort of deep learning increasingly used for other automatic target recognition, a trend driven by the ready availability of new, low-cost hardware for small uncrewed systems. (Hambling, 2021a, para. 2)

Only trickles of information about these weapon systems are available to civilians, but the available evidence often underscores that the technologies can be operated without direct human control and that the humans in charge may not be able to have direct roles in some forms of operations. Many of the systems are designed and produced by third-party outsourcers who generally consider detailed descriptions and safety test results as proprietary information (Yaacoub et al., 2022). Unfortunately, one of the major guides in public discourse for how robots and AI will operate in the military and security fields is still epic movies and science fiction, where the track record of the entities for being "situation aware" as well as merciful and emphatic is generally not very good.

Micro-level Control of Humans with Police and Security Robotics

With robots and AI as factors, intelligent and autonomous entities are already being introduced into comprehensible, everyday scenes and are altering how humans assess social contexts and make situational judgments. Awareness of the power of autonomous robots and military technologies is growing, which can enhance the deterrent capabilities of civilian applications (Scheutz & Malle, 2021). The following statement underscores the perceived attractiveness of such systems for military or civilian control efforts:

> The goal of developing AI-empowered battlefield robotic systems is to create an advantage over the adversary in terms of situational awareness, speed of decision-making, survivability of friendly forces, and destruction of adversary forces. Having unmanned robotic platforms bear the brunt of losses during combat—reducing loss of human life—while increasing the operational effectiveness of combat units composed of manned-unmanned teams is an attractive proposition for military leaders. (Swett et al., 2021)

The prospect of killer robots making life-and-death decisions in civilian and military contexts has stimulated discussion of how best to make these decisions, if they are required to be made at all (Kaplan, 2015).

Responses to these developments in civilian contexts have varied greatly as the potentials for "police robots" have expanded. Some commentators have welcomed these initiatives, largely for their capacities for relieving human officers and security guards of potentially dangerous duties (Szocik & Abylkasymova, 2021). In contrast, Pasquale (2020a) and others characterize the use of robotic systems for human supervision as "oppressive" (p. 27). Robotic dogs and police robots are already being purchased and implemented in some cities as part of crime-fighting measures; New York City's robotic police dog was labeled as "creepy" in numerous news and social media accounts (Offenhartz, 2021); it was eventually removed from operations. However, some other communities have retained their robots and critical questions about the character of the operations robots and autonomous entities provide have already emerged. For instance, Vincent (2017) described how an animal shelter in San Francisco used robots to scare off homeless people from the area in front of its establishment. In another example, Hawaii police used COVID-19 relief funds to buy a robotic dog for patrolling homeless encampments (Hignett, 2021, p. 1), and Winnipeg police have acquired one despite some citizen complaints that it was largely purchased as a "toy" (Robertson, 2021). Schellin et al. (2020) describe ongoing research that is designed to make the robot dogs more canine-like and effective in engaging users' attention. Robots have already killed individuals in civilian settings in the US by use of robot-wielded bombs (Glaser, 2016; Taft, 2016), and warfare conducted autonomously has already been a common occurrence internationally (Glaser, 2016). The American Civil Liberties Union (ACLU) expresses human rights concerns about these events in the following way: "It is easy to think up scenarios where weaponized robots might save the day, but such scenarios are likely to be rare—and meanwhile the potential for abuse and overuse is so significant that policymakers should closely monitor police departments' acquisition of and involvement with such machines" (Stanley, 2016). The prospects that such entities could be decoupled from their original programming through mishandling by novice users or by malicious manipulation via security breaches add to the concerns involved.

Descriptions of military and police robots include concerning if not devastating portrayals of punitive activity by robots directed to humans

(Young, 2021). The continuing stream of "deaths by robot" and other unfortunate acts described in this book are entering arenas that have already characterized robots in terms of their capabilities for violence. Robots, drones, and autonomous vehicles have been used for military and police purposes for decades, often in wartime theatres and developing nations in which the individuals being patrolled or otherwise controlled have little recourse (Casey-Maslen et al., 2018; Jackson, 2020; Swett et al., 2021). These technologies are rapidly becoming available to security agencies and police departments in various nations; as exemplified in an upcoming section, they are also being utilized by some private individuals with their own political or social agendas. Singapore has begun using robots for everyday patrols:

> Singapore started testing two robots, named "Xavier", to patrol public areas and prevent population abuse. In this first step, the following will be explored: Disregard for COVID-19 containment protocols; Cigarette use in a prohibited environment; and Bicycle parking at inappropriate places. Robots are equipped with cameras and trigger real-time alerts for a command and control center. The two "Xaviers" will initially operate in an area of high pedestrian traffic in the center of the country. (Wayne, 2021)

The potential for robots to take over some police, control, and military functions "may decrease dangers to police officers by removing them from potentially volatile situations" (Joh, 2016, p. 516). However, the potentials for "deskilling" of police also occur, with less acclimation to community environments and development of situation-specific know-how (Joh, 2019, p. 133). An example of the complex operations of security robots follows:

> Kansai Airport's two security robots will autonomously navigate and patrol routes, use a laser sensor to identify their locations and capture images with built-in cameras. Although the airport has its own security team, the introduction of the robots provides the airport with an extra layer of security. The robots allow for increased precision, power, effectiveness for tasks, increased safety when operating in hazardous areas, and can continue to operate without rests while carrying out repetitive tasks that humans may grow tiresome of. (Youd, 2021, para. 11)

Police and military robots provide new kinds of demonstrations of the capabilities of robots and AI in real-life social contexts; they also expose the perspectives and approaches of those who develop and implement them. The "robotic and AI imaginary" that shapes narratives about these technologies is being augmented with these accounts of "killer robots" and police "robodogs" that are designed to perform control functions in society, often autonomously. Knightscope's "rolling pickles" have reportedly had "few tangible results" in fighting crime, although they have been themselves the victims of crime as they are assailed by passers-by (Farivar, 2021). The supposed superiority of robots over humans in making certain kinds of decisions and actions is often being underscored in marketing and political initiatives in the effort to reinforce assumptions about the place of robots in the social order, including their roles in controlling human behavior.

The Dynamics of Fear Amplification

What was presented in the last few paragraphs is frightening enough. However, the dark side imageries and themes about robots and AI that are discussed throughout this book (derived from legacies of science fiction as well as real-life narratives) serve to amplify the power of killer robots and related technologies to influence societies and affect individual human beings in profound ways. These "killer" entities can help produce dread and anxiety in everyday life in the settings in which they are implemented (or are rumored to be implemented). In comparable ways, accounts of hackers stimulated enhanced "cybercrime" fears in previous decades, as "reporting of dystopic narratives about life in networked worlds shapes public reactions to technological change" (Wall, 2008, p. 861). Robotics and AI provide both a technological infrastructure and a trove of negative images and narratives that can be used in attempts to shape and even fine-tune how people think, feel, and behave. People can indeed fear various types of weaponry yielded by those in power, from Tasers to automatic rifles. However, something more sinister may be afoot as the entities evolve and acquire autonomous capacities: as stated in the *New York Times*, the "evolution of these [robotic] machines is considered a potentially seismic event in warfare, akin to the invention of gunpowder and nuclear bombs" (Satariano et al., 2021, para. 3). The changes in everyday environments and the basic assumptions of military and police activities often trigger "a shift in the trauma of civilians from a memory of the past to a perpetual anticipation of the threat of the future, subjecting increasing numbers of people to unending physical and psychological incarceration in a traumatising present" (Hoskins & Illingworth, 2020). Dissemination of the innovative new ways in which soldiers as well as civilians might die from autonomous weaponry is often part of their fear amplification potential. For instance, "a drone that gradually heated enemy soldiers to death" and "sonic weapons designed to wreck an enemy's hearing or balance" (Pasquale, 2020b, para. 5) conjure up mental imagery that can have long lasting psychological impacts, akin to those of medieval torture methods.

We cannot be sure as to the overall impact on human senses of security and wellbeing of having killer robots and related AI-enhanced entities as parts of our daily lives and social media narratives. Humans indeed share stories of specific horrific events, but they also communicate with each other their feelings of being unsettled by the "uncanny"—those emerging and vague narratives, images, and apparitions that are often associated with AI, robotics, and autonomous vehicles (Masschelein, 2011). Would a humanoid appearance affect how a military or security-enforcing machine is perceived? Human sensitivity to each other and the increasing rejection of discrimination based on physical characteristics may inspire individuals to be more accepting of robotic soldiers that have humanoid characteristics. (As related elsewhere in the book, in some cases the humanoid robots' characteristics could also make them appear more "creepy" and warn individuals against them, however.) Opportunistically capitalizing on humanity's various traits by using humanoid and canine forms for instituting robotic oppression is problematic, whether or not it would appear to be more humane to patrol individuals by using human-like entities.

Autonomous entities that are explicitly designed by various governmental authorities to have the power to make life-or-death decisions in military and police contexts can elicit considerable anxieties on the part of a populace. Despite the fact that some of the entities are designed to work on a "stealth," hidden basis, their very physical presence is often a factor in their effectiveness in creating fear and anxiety. Consider the following scenario of a weapon that can select and execute its targets without direct human intervention:

> Imagine an autonomous weapon which, in a fraction of a second, can scan a battlefield, pick out various potential targets, clearly determine whether or not they are armed, threatening, or dangerous, and then compare the individuals' features and likely identities against a compiled list of known enemy soldiers or insurgents. With this information in hand, the weapon

is then able to rapidly engage the individuals it determines are legitimate targets, and using its superior aiming abilities ensure that all munitions expended hit those targets without endangering anyone else. Before the first shell casing hits the pavement, all enemy contacts are neutralized, without a single civilian being harmed and without destruction of any property or risk to other friendly soldiers. (Wood, 2020, p. 223)

The selection of the "target" is generally within the discretionary power of the developers and implementers, though the mechanics of how the targets will be distinguished from other humans can be problematic. Should societal resources be invested in these attempts to develop robots and AI systems that deter humans from some sort of undesired behavior (or make them fearful enough to pull back from human or robot contact)? The engineering of fear in both civilian and military-related contexts through the development of intimidating-appearing technologies has been a considerable part of war and conflict research for decades (Van Creveld, 2010). Controversies about autonomous weaponry often underscore the indeterminacies of robotic and autonomous entity operations, increasing concern while not providing strategies for mitigation (Galliott & Wyatt, 2022). Leading figures in public policy have weighed in on these technical issues: for example, former US Secretary of State Henry Kissinger is quoted as stating that AI is "as consequential" but "less predictable" than nuclear weapons as a technology (Schizer, 2021).

The ready availability of advanced technology that can produce significant damage can empower criminal and terrorist initiatives as well as those linked to official government- or corporate-sponsored strategies, whether the terrorists are comprised of organizations, groups, or individuals. With the increasingly-available robotic and AI-enhanced tools, militias and even individuals can acquire some of the clout of armies and large organizations (Cronin, 2019). Consider the following example of how an individual's reported plans were to be realized through advanced technology:

> ... the adage that "life imitates art." Such was the case in September 2011, when Rezwan Ferdaus—a bright but disenchanted young man with a degree in physics from Northeastern University in Boston—was arrested for a plan to attack Washington, D.C., with high-end hobby remote-controlled aircraft. Ferdaus envisioned taking a squadron of F-86 Sabre and F-4 Phantom model planes filled with C-4 explosives to attack the U.S. Capitol and the Pentagon. They were to be launched from the East Potomac Park, near the Jefferson Memorial, and then guided by GPS to

their targets. Although many experts are skeptical that such hobby aircraft, which are approximately five feet long and capable of carrying a payload of about five pounds of C-4 plastic explosives, could do much against a target like the Pentagon, they certainly could demolish a car or motorcade. (Shaker, 2014, p. 6)

The fact that many advanced robotic and AI-enhanced vehicles are currently and potentially in the hands of terrorists presents the conditions for a high-tech "arms race" in which a society's research and development efforts in security and military technology eventually are used in criminal and anti-social manners. Members of a public who want to steer clear of these entities may be unsure of their origins and intents, and not knowledgeable of ways to extricate themselves from the situations involved.

CONTROL OF HUMANS BY ROBOTS AND AUTONOMOUS ENTITIES

This section explores the compulsory and even dominating nature of many implementations of robots and autonomous entities into human systems. A "slippery slope" syndrome is occurring as technologies that are accepted in military and workplace realms are being imported into wider applications and increasingly intimate matters. The notion that humans would be monitored in their everyday routines and their behavior controlled by robots in community or household settings may have seemed to be part of a dystopian and even bizarre science fiction fantasy just a few decades ago. However, robotic police and security entities are proliferating in many contexts, especially as pandemic-related or other kinds of public health assumptions and requirements have made direct human-to-human police work more difficult (Dellinger, 2021; Szocik & Abylkasymova, 2021). Some of these police and military entities are largely remote controlled but others are autonomous, and include land-based entities as well as drones (Enemark, 2021; Müller, 2020). "Swarm" robots with more diffuse presences are also being utilized (Olaronke et al., 2020). Attempting to shape human socialization so as to be more amenable to robot and AI control may be a lifelong endeavor, with children's "robot imaginaries" being formed early in their lives (Malinverni & Valero, 2020) and expanding to include various narratives and legacies as well as their household and workplace experiences with the technologies.

Police and security work is nearly always dangerous, and has added technical hazards when robots are involved; developers and implementers of robotic systems for police work have the challenge of considering the behavior of many individuals who do not have the goal of smooth human–robot interaction in mind. Images and narratives of robotic police dogs patrolling community spaces are beginning to fill research studies as well as tabloid press, serious journalism, and social media (Hignett, 2021; Robertson, 2021). Security robots equipped with facial recognition devices are already commonplace in community settings in a number of countries (Wayne, 2021; Youd, 2021). As these applications become normalized, the prospects for the deskilling of some forms of police work increase, as police perceive themselves as having less discretion when dealing with volatile situations. Confusions about the basic roles of the autonomous robots in relation to human roles are expanding as their capabilities increase (Cappuccio et al., 2021). Community participants will indeed receive some information about the police robots through the press, social media, and word-of-mouth, as well as whatever public information releases are provided by the relevant authorities. There will probably not be an "operations manual" for the police robots distributed to the public, which would leave humans guessing as to why the robots are engaging with them as they appear to be.

Some community settings are being redesigned to accommodate police and security robotics, placing the human even further into the "alien" mode in these environments (Miller, 1983). Increasingly, humans do not have a choice as to whether or not they interact with robots and AI, and may soon not have a choice as to whether they engage with autonomous vehicles in their daily lives. As robots are often being placed into dominant or superior positions, new kinds of tensions and related safety-related issues are emerging. For example, disturbing applications of robots and AI that have considerable public ramifications are developing as workplace and healthcare system administrators explore their increased potentials for targeted employee control. For example, some organizations are today utilizing evidence from robots and intelligence vehicles for legal purposes, analyzing the data collected for evidence that employees are inappropriately claiming workplace compensation for injuries on the job or otherwise misappropriating time and resources:

> … UPS [United Parcel Service in the US] trucks are functionally employee surveillance machines on wheels – they track everything from whether an

employee puts her seatbelt on before turning on the engine, to how many times a driver brakes in traffic, to how many minutes are "stolen" when the driver uses the bathroom. UPS workers report frequently being harassed by their bosses for a series of small actions recorded by these vehicles. (Mabud, 2019, para. 1)

Robots could even be used for entrapment, strategically inciting criminal activity; research by Aroyo et al. (2018) show how robots can potentially nudge individuals to "disclose sensitive information, conform to [their] recommendations, or gamble" (p. 3701). Through entrapment initiatives, individuals who possibly have tendencies toward committing a particular crime are motivated toward taking some recognizable step toward committing it (and are subsequently charged by authorities). Police efforts in many nations have come under increased scrutiny because of alleged abuses of entrapment; AI-enhanced predictive technologies could expand police capabilities in this regard (Oravec, 2022; Zaia, 2020). Police as human or robotic agents need to be identified, as well as need to have their scope of activity routinized, so that misunderstandings, miscommunications, and inappropriate entrapments do not occur.

INTERNATIONAL RESPONSES TO AUTONOMOUS MILITARY AND SECURITY SYSTEM PROSPECTS

Responses to killer robots have expanded as various nations and non-governmental organizations have sought transparency for their development and controls for their proliferation. The Campaign to Stop Killer Robots pioneered in fomenting opposition to the expansion of the use of autonomous entities by military forces (Carpenter, 2016; Gubrud, 2014). Nobel Peace Prize laureate Jody Williams along with Mary Wareham have been strong figures in the Campaign; their efforts were previously directed toward dealing with landmine proliferation. As of December 2021, the international community has run into serious obstacles in setting effective constraints for the proliferation of killer robots, with opposition by the US and several other countries to binding agreements:

> The US has rejected calls for a binding agreement regulating or banning the use of "killer robots", instead proposing a "code of conduct" at the United Nations. Speaking at a meeting in Geneva focused on finding common ground on the use of such so-called lethal autonomous weapons,

a US official balked at the idea of regulating their use through a "legally-binding instrument." ("US Rejects Calls," 2021, para. 1)

As of early 2022, no international agreement has been wrought. Statements such as the one below by the Vatican have attempted to focus attention on killer robot concerns in an era in which the pandemic has been the major source of anxiety:

> Vatican officials have joined Pope Francis in repeatedly expressing trepidation over such weapons, known as LAWS, saying their use poses a serious threat to innocent civilians. The most recent caution against so-called "killer robots" came from the Vatican Permanent Observer Mission to the U.N. agencies in Geneva in early August during a meeting of the 2021 Group of Government Experts on Lethal Autonomous Weapons Systems of the Convention on Certain Conventional Weapons. In the first of three daily statements to the group, the Vatican said Aug. 3 that a potential challenge was "the use of swarms of 'kamikaze' mini drones" and other advanced weaponry that utilize artificial intelligence in its targeting and attack modes. LAWS, the Vatican said, "raise potential serious implications for peace and stability." The statements were the most recent from Vatican diplomats at the U.N. in Geneva. Pope Francis also addressed autonomous weapons in his address to the 75th meeting of the U.N. General Assembly last September. In a world where multilateralism is eroding, he said, LAWS can "irreversibly alter the nature of warfare, detaching it further from human agency." (Sadowski, 2021)

The nonprofit Future of Life Institute (FLI), the International Committee of the Red Cross (ICRC), and various other groups contend that legally-binding prohibitions on the development of "autonomous weapons which use artificial intelligence to identify, select, and kill people without human intervention" are needed (Hambling, 2021b), much as how certain chemical and biological weapons are banned internationally.

As previously related, some individual researchers and developers have spoken out against these weapon systems (McDonald, 2019). Professional organizations have been slower to do so, despite the pleas of some activists:

> [Autonomous weapon systems] will permit armed conflict to be fought at a scale greater than ever, and at timescales faster than humans can comprehend. These can be weapons of terror, weapons that despots and terrorists use against innocent populations, and weapons hacked to behave

in undesirable ways. ... professionals should adopt a moratorium on the design of systems capable of using force autonomously, should ensure that human operators have meaningful control over any robotic system capable of deploying violent and lethal force against humans, and should consider permanently banning such systems in the future. (Asaro, 2016, p. 59)

At the end of the chapter, a brief history of the Computer Professionals for Social Responsibility (CPSR) is related to provide some directions as to how information technology professionals could organize to face these sociotechnical challenges today. Although the CPSR was disbanded a while back, its legacy for non-partisan collaboration and dissemination of expertise still endures.

Compulsory Robotics, Quartering, and the Third Amendment

The Pinkerton agents who surveilled US Ford Motor Company workers in their homes and communities many decades ago left an unfortunate legacy of corporate overreach (Weiss, 2014). Research and development initiatives to determine how best to increase the public's compliance to police and security robots may also demonstrate comparable forms of overreach. The compulsory exposure to robots and other AI-enhanced entities is raising issues of whether humans have a right not to have these entities in their immediate presence (or in their households). Legal and social protections are just emerging against the kinds of opportunistic behaviors of developers, distributors, and professional groups that are described in this book, including intrusive information collection as well as potentially-inappropriate control of environmental systems (such as a household's light and heat).

The creeping insinuation of robots and AI-enhanced technologies in nearly every part of our lives has inspired thinking about how to constrain the proliferations of these entities. In circumstances in which US governmental activity is involved, the Fourth Amendment of the US Constitution provides some legal protections against unwarranted searches and seizures, which have been generally interpreted by courts as encompassing computer technologies. The Third Amendment could also have some applicability as botnets and other intelligent agents are being effectively "quartered" in our homes just as flesh-and-blood human soldiers were in past centuries. Friedland (2014) relates that "continued quartering abuses in the colonies [by the British] led to the adoption of the Third Amendment" (para. 12). The Amendment, part of the US Bill of Rights, reads: "No Soldier shall, in time of peace be quartered in any house, without the consent of the Owner, nor in time of war, but in a manner to be prescribed by law." The late US Supreme Court Justice William O. Douglas used the Third Amendment in his efforts to uphold household privacy. Efforts to balance privacy with other compelling social and political ends could be difficult in robot- and AI-related cases as they have apparently been in other information technology and telecommunications settings.

A wide assortment of compulsory applications of robotics and AI are being integrated into everyday life in a manner that resembles a kind of "quartering": many of us are compelled in our households to have "smart" water and electricity meters, for example, that can often control our household water, heat, and light by central authorities and that can often require replacement when previous versions become technologically obsolete (Orlowski, 2022). Compulsory robotic interaction presents new issues as humans are also compelled to interact with security and service robots in their communities and workplaces. The potential for humans to extricate themselves from situations in which robots have a surveillance and control function is diminishing; for instance, as robots get smaller (such as the "swarm" robots described in Olaronke et al., 2020), the robots will be harder to detect, their purposes will be less transparent, they will be harder to defend against, and their "quartering" will be easier to accomplish (Oravec, 2017a).

SCIENCE FICTION AND AI-ENHANCED WARFARE AND CONTROL SYSTEMS

As outlined in previous chapters, individuals are learning about robotic and AI technologies through workplace efforts as well as various household and even intimate sexual initiatives. However, that some kinds of robots and AI-enhanced entities have the capabilities of controlling humans is a part of everyday acculturation, with many television shows, films, and comic books on these themes (Grupe, 1996), as well as continuing youth and adult exposure to these themes. This capacity to control has been extended in fiction to the capabilities to destroy humanity, with "killer robot" themes often blurring the actual capacities of our systems with the fictional projections of which we are inundated.

It would be impossible to ascertain just how extensive are the impacts of the images of robots from science fiction along with its characterizations of AI capabilities on today's military apparatus. Considering the cultures of militarization in advanced societies today is to discuss how the military is imbued into nearly every segment of society and how rules for participation as well as allowable resistance are established (Berland & Fitzpatrick, 2010). Even children are resented with these notions. For example, the movie Chappie (2015, Columbia Pictures) introduces young people to the notions and societal assumptions behind police robots: "as an introductory flurry of news reports indicates, law demands order. To help stem the murderous tide, the authorities have deployed a force of armed robots, called Scouts, to serve as shields for their human brethren or to fire on the heavily fortified shouting masses" (Dargis, 2015, para. 2). Military and police robots reinforce the image of robots as something that is not to be "turned off" by humans, with persistent and pervasive activity along with significant autonomy in their processes and decision-making.

Destruction by robots, drones, and autonomous vehicles presents significant perils for society, the fact that they are being normalized in light-hearted fashion in science fiction and gaming modalities intended for children is unsettling. Many BattleBot-style initiatives and robot clubs are rooted on the competitive dimensions of robots and violence, providing insights about today's society; robots, destruction, lack of transparency, and uncertainty are linked tightly, as they are in many movies and science fiction narratives (Sone, 2021). The witnessing of destruction of and by robots can have strong impacts on individuals over decades, normalizing technology-induced violence and increasing overall uncertainty and fear. This chapter emphasizes how uses of robotics and AI-enhanced entities for control of humans are becoming more feasible not just because of specific technological capabilities but because of how humans are increasingly demoralized, infantilized, and incapacitated by robots, autonomous vehicles, and the AI hype that accompanies them.

SOME CONCLUSIONS AND REFLECTIONS

As described in a previous chapter, many kinds of AI-enhanced entities besides those explicitly labeled as "killer robots" can indeed inflict lethal blows, for example through poor safety-related design or even because of the unpredictability of outcomes of the machine learning systems that guide their moves. However, military-commissioned killer robots are playing especially significant roles in civilian-level security discourse as well as international relations as rules for their proliferation are debated and as civilians explore the implications for their use. A large amount of military equipment of various sorts has migrated to local police and regional sheriffs in past decades (Bove & Gavrilova, 2017), so the eventual application of killer robot-style apparatus in everyday society is a strong possibility. Criminal and terrorist organizations are likely also to utilize these technologies in their own efforts to assert control over individuals, as equipment is either stolen or replicated by these forces. As previously discussed, a rationale that is often presented for the development of military and police robots and other AI-related entities is to relieve soldiers from the physical dangers of combat (including bodily casualties); however, the psychological toll on individuals who design and deploy autonomous military and security systems can also be considerable, especially as the foibles of these systems are exposed and innocent human beings lose their lives.

More than four decades ago, philosopher Langdon Winner asked "Do artifacts have politics?" (1980, p. 121). Today, the notion is widely accepted that a robot shaped like a dog, possessing various weaponry, and designed to intervene with humans in certain contexts does reflect certain political and social assumptions about the world. The notion of imposing control on a populace through lethal means without bodily risk to the human controller is taking root in many ways in modern societies. These assumptions are conveyed to people at relatively early ages: robots that are designed to battle each other in "robot competitions" until one is destroyed are part of many school activities (Crease, 2021), and use of some robots to patrol educational buildings is beginning. The basic notion of being a resident or citizen of a locale with such technological police installations is being formulated in terms of the requirements of autonomous robots as well: a major responsibility of community membership is to be in synch with the demands of the authorized police robots (however they are identified and certified). Just as the emergence of the personal computer shaped the parameters of many unprepared workplaces in the 1980s, the responsibility for using the technology in the best way possible can fall on ill-equipped and barely-knowledgeable users, with some police and security personnel not receiving adequate training with these advanced technologies. The technologies themselves are often not fully tested: as stated by Atanasoski and Vora (2020) "start-up culture has resulted in a world where new apps are massified in beta-test mode without any assessment of their social and ethical impact" (p. 1). The fact that outsourcers with government contracts are involved in the production and maintenance of these systems adds complexities to the situation since the responsibilities of these organizations in these emerging high-tech matters are often limited by contract or legal precedent, and transparency is often limited because of proprietary considerations (Calcara, 2021; Reis et al., 2021).

The foibles of robots in making poor decisions in highly-pressured and intense armed standoffs and protests has been foreseen in science fiction: "most literary works about robots are cautionary tales about insufficient programming, emergent behavior, errors, and other issues that make robots unpredictable and potentially dangerous" (Lin et al., 2011). Fear of police robots' unreliability may indeed affect the behavior of humans in proximity of the robots, but what may be even more disturbing is the potential for generalized, free-floating fear that can have deep cultural ramifications. Resolving these issues by increasing the amount of positive publicity concerning these military and security entities and classifying and hiding data that demonstrates their weaknesses is a short-term political solution to a much longer-term technosocial problem, but is an approach that is unfortunately all too common in this arena.

A Follow-Up: Deep Dive into the Computer Professionals for Social Responsibility (CPSR) of the 1980s and 1990s

[This section has some personal and historical reflections about a pioneering professional organization, no longer in existence, that stimulated discourse in past decades about the military and security issues of AI and computer networking. It is provided to stimulate thinking about what might be possible today.] The Computer Professionals for Social Responsibility (CPSR) had some differences with many other high-tech activist organizations in that it was not overtly political and refrained from endorsing candidates (Anderson, 1992; Finn & DuPont, 2020; Oravec, 2017b). I was a member early on as I (as a computer educator and researcher) struggled to understand how to do something about the social issues of computing. Some of our local Madison Wisconsin (US) CPSR members in the 1980s and 1990s did indeed want to make political stands with the use of CPSR endorsements, but they were voted down. CPSR chapters were fairly independent. Some CPSR members at our chapter worked quietly to support members who were hurt at

UW-Madison because of their political stances. (Unfortunately, academic freedom can be effectively challenged when grant and professional allocations are made.) The CPSR as a whole did not develop an ethical code, either, in part due to its respect for the differences of opinions and perspectives among members (Anderson, 1992). The CPSR was also not quite ready to mobilize into a labor union, although some members of our Madison chapter were clamoring for such a move. (Someone had worked in a union shop while doing computing research at Boeing Corporation.) However, the CPSR eventually did gain publicity and clout through taking strong and well-publicized stands, but against particular corporate products as well in response to specific current issues. For example, what gave CPSR some national attention in the early 90s was its protest against Lotus *Marketplace*, described in a previous section. The CPSR also participated in the dissemination of Y2K warnings, advising organizations as to how to handle the millennial date change, which had potential implications for some computer systems (I.P., 1998; Oravec, 1998). CPSR chapters sometimes disagreed with other computer-related groups about particular issues. For example, a CPSR chapter sent a letter to then President Bill Clinton declaring that its members disagreed with some statements by the Digital Privacy and Security working group (a coalition of other computer activist groups) concerning cryptography policy (Gurak, 1999).

In the early days of CPSR, a small percentage of the population could program and saying that programming was "stress-free" was very easy to do. I somewhat miss the days when computing expertise was relatively rare and I could obtain a sizable audience when giving a simple talk under the aegis of the CPSR about viruses and worms, as I did at Lawrence University in Appleton, Wisconsin ("Computers Used to Invade Privacy," 1988). My "backup television set" story provides an image of how the CPSR informed discussions of the societal risks of computing. The scenario transpired as follows: on 20 November 1983, the televised movie *The Day After* (ABC Network) provided a narrative of how the world would look in the event of widespread nuclear destruction. Our Madison, Wisconsin CPSR chapter met at the home of a member who demanded that we place a backup television on the top of the set that was our main focus for the event, just in case the primary TV set would break down. Some of us immediately began to chatter about risk and computing: what backup systems do we have that could prevent nuclear disaster? Could computer code be tested that would only be operable in extreme circumstances? And, do we really need a backup television set? Decades ago, a Cold War scenario was the focus, and China did not play the central role it has today as a high-tech powerhouse. What I hoped through the years did not transpire, however. CPSR was not able to play pivotal roles in decision-making about important technological policies and military applications. Although we as computer researchers could indeed speak on these matters in public arenas, what the world really wanted to hear were the utterances of generals and corporate leaders, few of whom had substantial computing expertise.

The backup TV set story also illustrates the multiplicity of the personalities and approaches of CPSR members. Much of the friendly banter during meeting breaks was about the eccentricities and exploits of members and colleagues. For example, we talked about an individual who had little spending money, so he ate three times a week at local all-you-can-eat places and fasted for the rest of the time. The kinds of rationalizations individuals made about their efforts to bend gastronomic rules in order to continue their research seemed oddly amusing then, but tremendously sad today. (Unfortunately, the individual did not last very long as a researcher.) We also discussed computing in terms of personal enhancements and extensions, bolstered by Marshall McLuhan's (1964) notions of technology as an extension of self along with early cyborg thinking. My presentation at a CPSR conference on "dependence on AI" (Oravec, 1988) and my book *Interactions in Science and Society* (Oravec, 1992), along with my legislative work on privacy were enlightened by these discussions. Technology had foibles but could also be empowering.

Who can effectively advise on complex matters concerning the use of computer research for military purposes? A bunch of often-emaciated computer researchers or military brass? President Reagan's "Star Wars" plans shocked many researchers who knew what advances would be needed to make automated missile defense a reality. CPSR's Boston contingent helped organize a debate related to the software reliability of Strategic Defense Initiatives (SDI) systems and obtained some national coverage (Boffey, 1986). The growing consensus that everyone can be a considered a computer expert (at least at some level) and be able to frame high-tech problems has somewhat taken the steam out of groups with lofty labels such as "Computer Professionals for Social Responsibility." The PSR, Physicians for Social Responsibility, has faced comparable issues about large-scale health issues. PSR began in 1961, but was dormant for a while until the late 1970s (Day & Waitzkin, 1985); like CPSR, it started with a heavy emphasis on nuclear-related themes then expanded its concerns more broadly to medical inequities and other public health concerns. The Electronic Frontier Foundation (EFF), initiated in 1990 by Mitch Kapor, John Perry Barlow, and John Gilmore, took on some of the issues and approaches of the CPSR (Boswell, 2021). Both the EFF and PSR still survive today, unlike CPSR. This brief reminiscence about CPSR was added here to provide some inspiration to those professionals and aspiring professionals who would like to design an organization of their own in order to tackle these pressing issues in a collaborative fashion.

REFERENCES

Anderson, R. E. (1992). Social impacts of computing: Codes of professional ethics. *Social Science Computer Review, 10*(4), 453–469.

Aroyo, A. M., Rea, F., Sandini, G., & Sciutti, A. (2018). Trust and social engineering in human robot interaction: Will a robot make you disclose sensitive information, conform to its recommendations or gamble? *IEEE Robotics and Automation Letters, 3*(4), 3701–3708.

Asaro, P. (2016). "Hands up, don't shoot!" HRI and the automation of police use of force. *Journal of Human-Robot Interaction, 5*(3), 55–69.

Atanasoski, N., & Vora, K. (2020). Why the sex robot becomes the killer robot: Reproduction, care, and the limits of refusal. *Spheres: Journal for Digital Cultures* (6), 1–16. https://mediarep.org/bitstream/handle/doc/14793/spheres_6_4-1_Atanasoski_Vora_Sex-Robot-Killer-Robot.pdf?sequence=1

Berland, J., & Fitzpatrick, B. (2010). Introduction: Cultures of militarization and the military-cultural complex. *TOPIA: Canadian Journal of Cultural Studies, 23*, 9–27.

Boffey, P. (1986, September 16). Software seen as obstacle in developing "Star Wars." *The New York Times*, p. 15. https://www.nytimes.com/1986/09/16/science/software-seen-as-obstacle-in-developing-star-wars.html

Boswell, S. F. (2021). "Jack in, young pioneer": Frontier politics, ecological entrapment, and the architecture of cyberspace. *American Literature, 93*(3), 417–444.

Bove, V., & Gavrilova, E. (2017). Police officer on the frontline or a soldier? The effect of police militarization on crime. *American Economic Journal: Economic Policy, 9*(3), 1–18.

Brown-Gaston, R. D., & Arora, A. S. (2021). War and peace: Ethical challenges and risks in military robotics. *International Journal of Intelligent Information Technologies (IJIIT), 17*(3), 1–12.

Calcara, A. (2021). Contractors or robots? Future warfare between privatization and automation. *Small Wars & Insurgencies, 33*, 250–271.

Cappuccio, M. L., Galliott, J. C., & Sandoval, E. B. (2021). Saving private robot: Risks and advantages of anthropomorphism in agent-soldier teams. *International Journal of Social Robotics*, 1–14. https://doi.org/10.1007/s12369-021-00755-z

Carpenter, C. (2016). Rethinking the political/-science-/fiction nexus: Global policy making and the campaign to stop killer robots. *Perspectives on Politics, 14*(1), 53–69.

Casey-Maslen, S., Homayounnejad, M., Stauffer, H., & Weizmann, N. (2018). Development, use, and transfer of unmanned weapons systems. In *Drones and Other unmanned weapons systems under international law* (pp. 8–45). Brill Nijhoff.

Chappie. (2015). Columbia Pictures. Directed by Neill Blomkamp.

Cheong, I. (2022. January 04). China deploys 'killer robots,' autonomous weapons platforms along Indian border. *Rebel News*. https://www.rebelnews.com/china_deploys_killer_robots_autonomous_weapons_platforms_along_indian_border

Computers used to invade privacy. (1988, November 29). *Post Crescent* (Appleton, WI), p. 22.

Crease, R. P. (2021). Combat robotics. *Physics World, 34*(5), 22.

Cronin, A. K. (2019). *Power to the people: How open technological innovation is arming tomorrow's terrorists*. Oxford University Press.

Dargis, M. (2015, March 5). Review: In 'Chappie,' a smart robot runs with the wrong crowd. *New York Times*. https://www.nytimes.com/2015/03/06/movies/review-chappie-a-smart-robot-in-a-violent-future.html

Day, B., & Waitzkin, H. (1985). The medical profession and nuclear war: A social history. *Journal of the American Medical Association (JAMA), 254*(5), 644–651.

Dellinger, A. J. (2021, June 14). Honolulu police spent $150k on a robot dog to monitor homeless people. *Mic*. https://www.mic.com/p/honolulu-police-spent-150k-on-a-robot-dog-to-monitor-homeless-people-81175598

Enemark, C. (2021). Armed drones and ethical policing: Risk, perception, and the tele-present officer. *Criminal Justice Ethics, 40*, 124–144.

Farivar, C. (2021, June 27). Security robots expand across U.S., with few tangible results. *NBCNews.com*. https://www.nbcnews.com/business/business-news/security-robots-expand-across-u-s-few-tangible-results-n1272421

Finn, M., & DuPont, Q. (2020). From closed world discourse to digital utopianism: The changing face of responsible computing at Computer Professionals for Social Responsibility (1981–1992). *Internet Histories, 4*(1), 6–31.

Friedland, S. I. (2014). The Third Amendment, privacy, and mass surveillance. *Wake Forest Law Review Online*, 4. http://www.wakeforestlawreview.com/2014/02/the-third-amendment-privacy-and-mass-surveillance/

Galliott, J., & Wyatt, A. (2022). A consideration of how emerging military leaders perceive themes in the autonomous weapon system discourse. *Defence Studies*, 22(2), 253–276.

Glaser, A. (2016, July 11). 11 police robots patrolling around the world. *Wired*. https://www.wired.com/2016/07/11-police-robots-patrolling-around-world/

Grupe, F. H. (1996). In our own image: Robots in cartoons. *Journal of Systems Management*, 47(1), 58–63.

Gubrud, M. (2014). Stopping killer robots. *Bulletin of the Atomic Scientists*, 70(1), 32–42.

Gurak, L. J. (1999). The promise and the peril of social action in cyberspace. In *Communities in cyberspace* (pp. 243–263). Routledge.

Hambling, D. (2021a, November 4). Australian Army getting bulletproof swarming attack robots. *Forbes*. https://www.forbes.com/sites/davidhambling/2021/11/04/australian-army-gets-bulletproof-attack-robots/?sh=42222f00463b

Hambling, D. (2021b, December 3). 'If human, kill': Video warns of need for legal controls on killer robots. *Forbes*. https://www.forbes.com/sites/davidhambling/2021/12/03/new-slaughterbots-video-warns-of-need-for-legal-controls-on-killer-robots/?sh=17565a927238

Hignett, K. (2021, June 18). Hawaii police used Covid relief funds to buy a robotic dog for patrolling homeless camp. *Metro.co*. https://metro.co.uk/2021/06/18/us-police-bought-a-robot-dog-to-patrol-homeless-a-camp-14792546/

Hoskins, A., & Illingworth, S. (2020). Inaccessible war: Media, memory, trauma and the blueprint. *Digital War*, 1, 74–82.

I.P. (1998). Millennium bug bites? *Science News*, 153(3), 45.

Jackson, R. D. (2020). "I approved it... and I'll do it again": Robotic policing and its potential for increasing excessive force. In *Societal challenges in the smart society* (pp. 511–522). Universidad de La Rioja.

Joh, E. E. (2016). Policing police robots. *UCLA Law Review Discourse*, 64, 516.

Joh, E. E. (2019). The consequences of automating and deskilling the police. *UCLA Law Review Discourse*, 67, 133.

Kaplan, J. (2015, August 17). Robot weapons: What's the harm? *New York Times*, A19.

Kelly, F. N. (2018). Analyzing the potential for universal disarmament of autonomous weapons systems or how I learned to stop worrying and love the killer robot. *Brookings Journal International Law*, 44, 381–409.

Lin, P., Abney, K., & Bekey, G. (2011). Robot ethics: Mapping the issues for a mechanized world. *Artificial Intelligence, 175*(5–6), 942–949.

Mabud, R. (2019, January 22). When the real threat is worker surveillance—Not the robot apocalypse. *Forbes*. https://www.forbes.com/sites/rakeenmabud/2019/01/22/when-the-real-threat-is-worker-surveillance-not-the-robot-apocalypse/?sh=572ded0f6a2f

McDonald, H. (2019, September 15). Ex-Google worker fears 'killer robots' could cause mass atrocities. *The Guardian*. https://www.theguardian.com/technology/2019/sep/15/ex-google-worker-fears-killer-robots-cause-mass-atrocities

Malinverni, L., & Valero, C. (2020, June). What is a robot? An artistic approach to understand children's imaginaries about robots. In *Proceedings of the Interaction Design and Children Conference* (pp. 250–261).

Masschelein, A. (2011). *The unconcept: The Freudian uncanny in late-twentieth-century theory*. Suny Press.

McLuhan, M. (1964). *Understanding media: The extensions of man*. MIT Press.

Miller, R. J. (1983). The human: Alien in the robotic environment? *The Annals of the American Academy of Political and Social Science, 470*(1), 11–15.

Müller, O. (2020). "An eye turned into a weapon": A philosophical investigation of remote controlled, automated, and autonomous drone warfare. *Philosophy & Technology, 34*, 875–896.

Offenhartz, J. (2021, April 13). NYPD deploys "creepy" new robot dog in Manhattan public housing complex. *Gothamist*. https://gothamist.com/news/nypd-deploys-alarming-robot-dog-manhattan-public-housing-complex

Olaronke, I., Rhoda, I., Gambo, I., Oluwaseun, O., & Janet, O. (2020). A systematic review of swarm robots. *Current Journal of Applied Science and Technology, 39*, 79–97.

Oravec, J. A. (1988). Dependence upon expert systems: The dangers of the computer as an intellectual crutch. In *Directions and Implications of Advanced Computing Symposium Proceedings*, Computer Professionals for Social Responsibility (CPSR). University of Minnesota.

Oravec, J. A. (1992). *Interactions in science and society*. Agency for Instructional Technology.

Oravec, J. A. (1998). Learning from Y2K: The Y2K problem has made managers rethink the role of IT. *Ivey Business Journal, 63*(2), 20–25.

Oravec, J. A. (2017a). Kill switches, remote deletion, and intelligent agents: Framing everyday household cybersecurity in the internet of things. *Technology in Society, 51*, 189–198.

Oravec, J. A. (2017b). Experiments in protesting: Virtual sit-ins, hacktivism, electronic civil disobedience, and other technology-mediated protest before the advent of social media. In *'Voices of Dissent': Social Movements and Political Protest in Post-war America*. Rothermere American Institute,

University of Oxford. https://www.rai.ox.ac.uk/sites/default/.../voices_of_dissent_programme_12.5.17.pdf

Oravec, J. A. (2022). The emergence of "truth machines"?: Artificial intelligence approaches to lie detection. *Ethics and Information Technology, 24*(1), 1–10.

Orlowski, A. (2022, January 3). Careless civil servants are about to cause chaos with smart meters: A determination to turn off 2G means that millions of the devices will have to be replaced—Again. *The Telegraph*. https://www.telegraph.co.uk/business/2022/01/03/careless-civil-servants-cause-chaos-smart-meters/

Osborn, K. (2021, May 31). America's military wants a robot Army. *National Interest*. https://nationalinterest.org/blog/buzz/america%E2%80%99s-military-wants-robot-army-186371

Pasquale, F. (2020a). *New laws of robotics: Defending human expertise in the age of AI*. Harvard University Press.

Pasquale, F. (2020b). 'Machines set loose to slaughter': The dangerous rise of military AI. *The Guardian*. https://www.theguardian.com/news/2020/oct/15/dangerous-rise-of-military-ai-drone-swarm-autonomous-weapons

Ploumis, M. (2022). AI weapon systems in future war operations; strategy, operations and tactics. *Comparative Strategy, 41*(1), 1–18. https://doi.org/10.1080/01495933.2021.2017739

Reis, J., Cohen, Y., Melão, N., Costa, J., & Jorge, D. (2021). High-tech defense industries: Developing autonomous intelligent systems. *Applied Sciences, 11*(11), 4920.

Reynolds, G. H. (1991). Outer space and peace: Some thoughts on structures and relations. *Tennessee Law Review, 59*, 723.

Rezende, I. N. (2020). Facial recognition in police hands: Assessing the 'clearview case' from a European perspective. *New Journal of European Criminal Law, 11*(3), 375–389.

Robertson, D. (2021, July 28). Winnipeg police acquire controversial robot dog. *Winnipeg Free Press*. https://www.winnipegfreepress.com/local/winnipeg-police-acquire-controversial-robot-dog-574954652.html

Rosert, E., & Sauer, F. (2021). How (not) to stop the killer robots: A comparative analysis of humanitarian disarmament campaign strategies. *Contemporary Security Policy, 42*(1), 4–29.

Sadowski, D. (2021, August 26). Push to ban 'killer robots' gets boost from Vatican, Pope. *Catholic Review*. https://catholicreview.org/push-to-ban-killer-robots-gets-a-boost-from-the-vatican-and-the-pope/

Satariano, A., Cumming-Bruce, N., & Gladstone, R. (2021, December 17). Killer robots aren't science fiction. A push to ban them is growing. *New York Times*. https://www.nytimes.com/2021/12/17/world/robot-drone-ban.html

Schellin, H., Oberley, T., Patterson, K., Kim, B., Haring, K. S., Tossell, C. C., Phillips, E., & de Visser, E. J. (2020, April). Man's new best friend? Strengthening human-robot dog bonding by enhancing the doglikeness of Sony's Aibo. In *2020 Systems and Information Engineering Design Symposium (SIEDS)* (pp. 1–6). IEEE.

Scheutz, M., & Malle, B. F. (2021). May machines take lives to save lives? Human perceptions of autonomous robots (with the capacity to kill). *Lethal Autonomous Weapons: Re-Examining the Law and Ethics of Robotic Warfare.*

Schizer, M. W. (2021, November 12). Henry Kissinger says AI is "as consequential" but "less predictable" than nuclear weapons. *Newsweek.* https://www.newsweek.com/2021/11/12/henry-kissinger-says-ai-consequential-less-predictable-nuclear-weapons-1644508.html

Shaker, S. M. (2014). Good robots gone bad. *The Futurist, 48*(3), 6.

Sharkey, A. (2019). Autonomous weapons systems, killer robots and human dignity. *Ethics and Information Technology, 21*(2), 75–87.

Sone, Y. (2021). BattleBots, machine surrogates and the organology of violent televisual entertainment. In *Distributed perception* (pp. 247–260). Routledge.

Stanley, J. (2016, December 20). The use of killer robots by police. *American Civil Liberties Union (ACLU).* https://www.aclu.org/blog/criminal-law-reform/reforming-police/use-killer-robots-police

Swett, B. A., Hahn, E. N., & Llorens, A. J. (2021). Designing robots for the battlefield: State of the art. In J. von Braun, M. Archer, G. M. Reichberg, & M. Sánchez Sorondo (Eds.), *Robotics, AI, and humanity.* Springer. https://doi.org/10.1007/978-3-030-54173-6_11

Szocik, K., & Abylkasymova, R. (2021). Ethical issues in police robots: The case of crowd control robots in a pandemic. *Journal of Applied Security Research,* 1–16. https://doi.org/10.1080/19361610.2021.1923365

Taft, I. (2016, July 8). Police use of robot to kill Dallas suspect unprecedented, experts say. *Texas Tribune.* https://www.texastribune.org/2016/07/08/use-robot-kill-dallas-suspect-first-experts-say/

US rejects calls for regulating or banning 'killer robots' (2021, December 2). *The Guardian.* https://www.theguardian.com/us-news/2021/dec/02/us-rejects-calls-regulating-banning-killer-robots

Van Creveld, M. (2010). *Technology and war: From 2000 BC to the present.* Simon and Schuster.

Vincent, J. (2017, December 13) Animal shelter faces backlash after using robot to scare off homeless people. *The Verge.* https://www.theverge.com/2017/12/13/16771148/robot-security-guardscares-homeless-san-francisco

Wall, D. S. (2008). Cybercrime and the culture of fear: Social science fiction(s) and the production of knowledge about cybercrime. *Information, Communication & Society, 11*(6), 861–884.

Wayne, M. (2021, September 7). Singapore tests 'Xavier', robot that detects population abuse. *Sprout Wired*. https://www.sproutwired.com/singapore-tests-xavier-robot-that-detects-population-abuse-world/

Weber, J. (2021). Artificial intelligence and the socio-technical imaginary: On Skynet, self-healing swarms and Slaughterbots. In *Drone imaginaries* (pp. 167–179). Manchester University Press.

Weiss, R. (2014). Corporate security at Ford Motor Company: From the Great War to the Cold War. In *Corporate security in the 21st century* (pp. 17–38). Palgrave Macmillan.

Winner, L. (1980). Do artifacts have politics? *Daedalus, 109*, 121–136.

Wittes, B., & Blum, G. (2015). *The future of violence: Robots and germs, hackers and drones-confronting a new age of threat*. Basic Books.

Wood, N. G. (2020). The problem with killer robots. *Journal of Military Ethics, 19*(3), 220–240.

Yaacoub, J. P. A., Noura, H. N., Salman, O., & Chehab, A. (2022). Advanced digital forensics and anti-digital forensics for IoT systems: Techniques, limitations and recommendations. *Internet of Things*, 19. https://doi.org/10.1016/j.iot.2022.100544

Young, G. (2021). On the indignity of killer robots. *Ethics and Information Technology, 23*, 473–482.

Youd, F. (2021, November 25). Do the robot: Kansai Airport's new autonomous security. *Airport Technology*. https://www.airport-technology.com/features/do-the-robot-kansai-airports-new-autonomous-security/

Zaia, M. (2020). Forecasting crime? Algorithmic prediction and the doctrine of police entrapment. *Canadian Journal of Law and Technology, 18*(2). https://doi.org/10.2139/ssrn.3644458

CHAPTER 6

Gilding Artificial Lilies: Artificial Intelligence's Legacies of Technological Overstatement, Embellishment, and Hyperbole

The chapter explores how overstatements and hyperbolic themes and concepts, often stemming from AI's early periods, are being employed in characterizations of current AI approaches in apparently opportunistic attempts to provide rhetorical support for various large-scale business and societal initiatives as well as consumer products. It also addresses the relative neglect of the consideration of many of AI's sociotechnical failures and discontinued approaches in recent examinations of automation and human welfare issues. These unfortunate patterns from decades past portend comparable kinds of issues today as robotics, automated vehicles, and other AI-enhanced entities enter everyday life; we are entering an era of "robo-hype" as developers and marketers present robotic features in overstated terms. The chapter discusses the moral logic of AI researchers and developers providing reasonable and measured narratives in public discourse rather than hyperbole, efforts that can empower decision makers to make sounder judgments concerning AI's current and projected applications (such as autonomous vehicles and robots) as well as allocate the rewards of the technology more equitably.

Much of the "dark side" discourse of AI since the 1950s has involved job loss (Rifkin, 1995). AI technologies are often infused into system development efforts that have resulted in considerable job losses and dislocations. These displacements have frequently been framed by developers, implementers, and technology advocates in ways that construe the disruptions as inevitable and often even desirable, rooted in constructed historical tapestries in which past and current levels of progress are inflated and failures are often downplayed (Pankewitz, 2017; Wacjman, 2017). Overstatements of technological progress and even inserts of hyperbole may be an expected part of grant proposals or research initiative descriptions, as provided for the US's "Strategic Defense Initiative" (SDI) in the 1980s (Van Nostrand, 2013) and some aspects of genetic engineering and biology (Caulfield, 2000; Coleman, 2016), as well as for information technology initiatives (Oravec & Travis, 1992; Ramiller, 2006). However, the insertion of such overstatements into the publically-disseminated characterizations of technological advances may have direct significance for the lives of individuals as well as have considerable ethical dimensions. For example, many employees understand little about the details of the AI systems involved in their workplaces but are still profoundly affected by how the systems and their outcomes are characterized (Forsythe, 2001; Powles & Hodson, 2017). Some pharmaceutical and medical marketing initiatives in critical care realms have faced comparable ethical issues in framing products so that consumer interest is attracted while providing reasonable levels of hope for positive outcomes (Ostergren et al., 2015). Overstatements and hyperbole concerning AI and its technological products can engender squandered resources and false hopes on the part of system developers as well as anxieties by those currently or potentially affected by AI implementations.

Technological Overstatement and Hyperbole

This section provides some historical background on artificial intelligence (AI) and analyzes a selection of writings and public statements produced since the 1940s concerning its emerging economic and social issues. Some of these statements provide warnings about potential challenges in the future regarding the use of robotics and AI; others present a more enthusiastic portrait of where society is headed. Overstatement in technological realms can indeed be inadvertent, related to lack of knowledge and changing conditions; the production of hyperbole often has some conscious intent, however. From a dramaturgical and rhetorical perspective, hype embellishes and exaggerates factual statements for specific purpose; it provides an "augmentation of real states of affairs" (Nemesi,

2004, p. 353). Szathmári (1958) explains hyperbole by contending that there is indeed another, more adequate and recognizable way to convey the notions outlined in the hype, but the speaker or author wants to provide the embellishment for some specific effect (p. 146). In other words, hyperbole generally incorporates some sort of conscious element to its overstatement of magnitudes or dimensions. Hyperbole is often "used to garner public interest and support for science" (Coleman, 2016, p. 1), playing special roles when competition for limited research resources is a factor. In its positive formulation, AI hype is "based on high positive expectations towards AI contribution to economic growth and addressing societal problems" (Ulnicane et al., 2021); from a more negative and cynical perspective, AI hype consists of fabrications that enable researchers and developers to promote AI without producing its promised outcomes. AI has been associated with dramatic levels of hyperbole throughout its history, with an upsurge in the past decade (Wajcman, 2017). Selections of what historical materials about origins and development of technologies to incorporate into public policy and marketing statements can be a part of strategic hyperbolic efforts; characterizations of these writings and quotations can also send signals about how the technologies are being framed for various audiences. For example, frequent references in current writings to the work of Herbert Simon, Norbert Wiener, Marvin Minsky, and contemporaries from previous decades have emerged (Hollnagel, 2014), linking AI to strong and evocative declarations about technological futures without requiring its advocates to compose new and possibly problematic statements.

The kinds of notions reflected in AI-related public discourse need to be chosen carefully so that the trends and nuances of today's research and development efforts can be more adequately treated. The promotion and subsequent dismantling of AI hype has been regular enough to construct "AI hype cycles" that inflate and deflate through time (Hughes & Hughes, 2019). The proliferation of AI hype is not without negative consequences: hype can have specific implications for how individuals behave and in their plans for the future. For example, Banja (2020) describes how statements of AI hype about computers replacing trained radiologists are affecting the field: they are "frankly irresponsible, at least to the extent that they discourage promising medical students from careers in radiology" (p. 1). Liu wrote "it may be that the dazzle of AI (as with other emerging technologies) in the context of existential risks is crippling our ability to respond" (p. 25). Computer science students also need to be reminded of AI's origins and limitations: pioneering AI scientist Fei-Fei Li stated "I often tell my students not to be misled by the name 'artificial intelligence'—there is nothing artificial about it. AI is made by humans, intended to behave by humans, and, ultimately, to impact humans' lives and human society" (quoted in Metz, 2018).

Discourse on artificial intelligence (AI) has often been contentious and argumentative, with hype playing considerable roles: in the past few years, a number of public policy debates and vocal protests specifically directed at AI and automation developers have arisen, both in high-tech-oriented communities as well as primarily industrialized regions (Akst, 2013). Many of these controversies have focused on such technologies as drones, robots, self-driving cars, and online transportation services such as Uber (Brynjolfsson & McAfee, 2011; Ranchordás, 2017; Richardson, 2015; Vogel, 2017). These interactions as well as extensive social commentaries and public policy initiatives have signaled some disquiet with the ways that human capacities are being construed in relation to those of machines, primarily in the practical ways in which AI systems are often given preference over human know-how in certain workplace and community settings. As described in this chapter, many expressions from AI developers as well as marketing and public policy advocates are stressing the positive aspects of AI, often omitting or underplaying the failures, setbacks, and tradeoffs described in historical treatments as well as recent accounts. This chapter has the limitation of focusing on US and UK approaches to these technologies and related controversies; however, there have indeed been some substantial contributions to AI research from other nations, generally with less publicity and self-promotion (Bruderer, 2017).

AI in Historical Perspective

The origins of AI are often linked to pioneering research presented by John McCarthy, Arthur Samuel, Oliver Selfridge, Allen Newell, and others at a conference at Dartmouth University in 1956 (Cordeschi, 2007). The name "artificial intelligence" reportedly emerged at this conference. An excerpt from the conference's ambitious grant proposal is below:

> We propose that a 2 month, 10 man study of artificial intelligence be carried out during the summer of 1956 at Dartmouth College in Hanover,

New Hampshire. The study is to proceed on the basis of the conjecture that every aspect of learning or any other feature of intelligence can in principle be so precisely described that a machine can be made to simulate it. An attempt will be made to find how to make machines use language, form abstractions and concepts, solve kinds of problems now reserved for humans, and improve themselves. We think that a significant advance can be made in one or more of these problems if a carefully selected group of scientists work on it together for a summer. (McCarthy et al., 1955, para. 1)

The notion that a summer would be enough time to make significant progress on issues that are still consuming the energies and resources of many scientists indeed underscores the optimism and even hubris of these AI pioneers. Another AI pioneer, Nobel laureate Herbert Simon, characterized the early momentum of AI as being kindled by an interest in the "nature of intelligence":

AI deals with some of the phenomena surrounding computers, hence is a part of computer science... It is also a part of psychology and cognitive science. It deals, in particular, with the phenomena that appear when computers perform tasks that, if performed by people, would be regarded as requiring intelligence-thinking. Artificial intelligence began in the 1950s as an inquiry into the nature of intelligence. (Simon, 1995, p. 95)

Some of AI's theoretical and mathematical underpinnings were provided by the UK's Alan Turing in previous decades (Turing, 1950). Many of the early AI research efforts were linked to larger military research initiatives; in the US, DARPA (Defense Advanced Projects Research Agency, later known as ARPA, Advanced Research Projects Agency) provided a great deal of support during this era (Van Nostrand, 2013).

Computer science as a whole, a very young discipline, began to blossom in the 1950s and 1960s; clustering a number of methodologies around a technology such as a computer seemed strange to many in the academic community who were used to working in more traditional disciplines such as mathematics. In the course of various projects, AI researchers sometimes discovered that very useful work had already been done by the psychology, linguistics, library science, and archiving communities, work that was often highly relevant to the systems development they were undertaking (Olson, 2001). Demarcations between "real" computer science and the efforts of those in other fields were often hastily drawn and turf battles continued through the decades as various academic units morphed into "information science" or "information studies" departments (Forsythe, 2001). Varieties of AI research such as "perceptrons" and neural networks (often linked to the work of Frank Rosenblatt) also began to form during these early days but did not acquire the research funding and other academic support of more cognitive and logic-oriented approaches (Rosenblatt, 1958). Neural network approaches inspired by Rosenblatt's and related efforts subsequently blossomed in the 2000s; massive machine learning research initiatives were eventually sponsored by major international corporations such as Google (now part of Alphabet Corporation), Amazon, IBM, Facebook, Apple, and others.

"Cybernetics" is an AI-related perspective with strong roots in the 1940s; as a term it is less often used today but its approach is still linked to many particular research and development initiatives (Kline, 2015). In the 1948 book *Cybernetics,* Norbert Wiener defined it in the following way: "Cybernetics is the study of communication and control in animals and machines, communication being the receiving and digesting of information, and control the use of this information in a direct action" (p. 1). The notion of humans, animals, and machines having comparable sorts of control and communication structures may seem familiar today but triggered controversy and concern in the 1940s and 1950s (Markoff, 2013). According to Hong (2004), Wiener's thinking "symbolized the beginning of a new conception of the man–machine relationship" (p. 50). Although artificial intelligence was still in its infancy in the 1950s, projections of substantial job loss and related economic stresses proliferated in academic literature and popular press accounts (Armstrong et al., 2014); many of the anti-automation protests that followed in the 1960s were linked to military research efforts, which provided considerable grounding for cybernetics and AI projects.

The ambitious notion of the "thinking machine" has been associated with AI research and development since its beginnings, often providing an aspirational component to various research and development agendas. Such machines were generally construed as autonomous intelligent entities independently operating on the world with various degrees of autonomy and self-awareness (Heilbroner, 1965; Hurlburt, 2017). Many futuristic projections emerged through the decades of how these entities (including chatbots as well as more sophisticated and extensive autonomous systems) could change everyday life (Johnson & Verdicchio, 2017). Thinking machines have played roles in science fiction novels and other entertainment genres, often in the form of intelligent and mobile robots. Some scenarios of the future project that humans themselves will be displaced in their basic functions by intelligent systems (Heylighen & Lenartowicz, 2017; Simpson, 2016). Battles between humans and machines for contested social spaces have often been projected, with the terrains often constructed in ways that give AI systems an inherent advantage. AI research has often been framed as experiencing successes in particular contest and gaming confrontations in the past decades: for example, chess-playing computers can defeat grand masters, as predicted by a number of AI pioneers (Arnold & Scheutz, 2016; Miller, 1992), and AI systems such as IBM's *Watson* have taken on celebrity status in part because of particular successes in US game shows such as *Jeopardy* (Lee & Kim, 2016). The thinking machine notion may indeed seem simple enough to be understood by a general audience but contains many assumptions and requires research and development underpinnings that involve advanced technological support. Technological narratives impart structures for dealing with technologies and their impacts in household, workplace, and public policy venues (van der Laan, 2016). By framing AI in generally positive terms that downplay failures and setbacks, individuals can be disempowered who need to make sound decisions about AI systems deployment. For example, Newlands (2021) describes how some AI service providers, in order to put their processes into operation in an organization, "must 'lift the curtain,' resulting in a paradoxical situation of needing to both perpetuate dominant AI hype narratives while emphasising AI's mundane limitations" (p. 2).

For the past decades, the prospects for AI applications to lead to societal disruption were often a part of societal discourse: for example, Lehman-Wilzig (1981) characterized the AI advances of the 1950s through the 1970s as "Frankenstein unbound." Norbert Wiener's 1950 work *The Human Use of Human Beings* (republished as Wiener, 1988) posited a series of difficult questions about the future of human intellectual efforts; he claimed that we were entering "a special period in the history of the world" (p. 46) in which shifts in technology result in massive societal alterations. Conway and Siegelman (2004) identify Wiener as the "dark hero" of the information age for his intense focus on the potentially-negative ramifications of AI technologies (p. 1). The image of the robot replacing human workers was linked to many cartoons, television shows, movies, editorial commentaries, and related expressions (Johnston, 2008). Herbert Simon reportedly stated that "there are now in the world machines that think, that learn, and that create. Moreover, their ability to do these things is going to increase rapidly until in a visible future the range of problems they can handle will be coextensive with the range to which the human mind has been applied" (from Simon & Newell, 1958). Detailed philosophical perspectives on the relative roles and statuses of humans and machines in society were crystalized and disseminated in the 1950s and 1960s (Taube, 1963), and many humanists and economists as well as technical experts participated in the discussions of this early period (as related in Bolisani, 2008; Kling, 1991; and Noble, 1995). An assortment of economic analyses rooted in these technological changes subsequently followed: for example, Marc Porat's (1977) identification and cataloguing of various knowledge professions stimulated interest in the notion of the "knowledge economy" and awareness of its economic dimensions (Hellström & Raman, 2001).

Herbert Simon's cognitive and logic-based approaches to AI dominated for a great deal of 1950s through the 1970s (Langlois, 2003). In *The Sciences of the Artificial* (1996/1961), Simon addresses human capabilities with the following: "Human beings, viewed as behaving systems, are quite simple. The apparent complexity of our behavior over time is largely a reflection of the complexity of the environment in which we find ourselves." Norbert Wiener develops some comparable themes contrasting human and computer capabilities: "As machines learn they may develop unforeseen strategies at rates that baffle their programmers.... By the very slowness of our human actions, our effective control of machines may be nullified. By the time we are able to react to information conveyed to our senses and stop the car we are driving, it may already have run head on into a wall.... Therefore, we must always exert the full strength of our imagination to examine where the full use of our new modalities may lead us" (Wiener, 1960, p. 1355). In this early period, AI-related discourse began to tackle the issue of whether autonomous systems are capable of moral reasoning (Arnold & Scheutz, 2016), thus supposedly minimizing the needs for human control and ethical judgment in certain sensitive, critical situations. In the 1980s, the book *The Fifth Generation* popularized a new set of AI approaches linked to "knowledge engineering" and "expert systems," drawing increased levels of corporate and military support (Feigenbaum & McCorduck, 1984). The paperback

The First Artificial Intelligence Coloring Book was intended to demonstrate the supposed clarity and simplicity of the notions underpinning AI systems (Cohen et al., 1983); supposedly, even young people could understand the basic ideas behind the technology involved. Implementing AI applications was apparently more difficult than projected in the comic book, which inspired some of the failure-themed narratives provided in a later section.

CHARACTERIZING ARTIFICIAL INTELLIGENCE

Why should efforts to characterize AI or other technologies be of concern outside of research contexts? Individuals who are reportedly losing their jobs to entities labeled as "intelligent technologies" or "AI systems" may indeed request some sort of account of what comprises the entities involved. Public policy researchers, community organizers, and business leaders may also seek such accounts to aid in their decision making. Definitions of AI have been sought for decades, ways of encapsulating the many research efforts that have been placed under that rubric, including logic programming, machine learning, natural language processing, expert systems, and many more (Lehman-Wilzig, 1981). The "Turing Test" (based on the "imitation game" of decades past) has often been used as a way to focus discussion on what constitutes an intelligent machine (Saygin et al., 2000; Woolgar, 1989). The Turing Test is itself a kind of contest, pitting various computer systems against each other as to which could outsmart humans most effectively as to whether or not it is itself a human. Aspects of AI research have varied through the past decades, with particular approaches such as object-oriented programming becoming more "mainstream" and thus not fitting under the AI rubric (Crevier, 1993). The ethical responsibility of AI researchers and developers to describe their initiatives in ways that foster informed deliberation about AI's prospects and build appropriate levels of legitimacy for its implementations is growing as the range of AI applications expands. As described in this chapter, more nuanced and measured narratives could help in reframing AI (along with robotics and autonomous vehicle research) so that the society as whole that has invested so much in AI's development can reap appropriate levels of rewards from its applications.

Although narratives expressing disquiet about the impacts of AI are indeed being produced, so are many highly optimistic characterizations (with descriptions such as "breathtaking"). This optimism is often related to AI's potential financial gains:

> The field of artificial intelligence is, indeed, making breathtaking advances. In particular, it is contributing to the automation of data analysis. Artificial intelligence is no longer programmed line by line, but is now capable of learning, thereby continuously developing itself. Recently, Google's DeepMind algorithm taught itself how to win 49 Atari games. Algorithms can now recognize handwritten language and patterns almost as well as humans and even complete some tasks better than them. They are able to describe the contents of photos and videos. Today 70% of all financial transactions are performed by algorithms. News content is, in part, automatically generated. (Helbing et al., 2017, p. 2)

Morris et al. (2017) write of the "remarkable resurgence" of AI research and development, describing its increasing use in automation; Bollier (2017) declares that AI has come of age" (p. 1). A special report in *The Economist* relates that "After many false starts artificial intelligence (AI) has taken off" but provides little background on these past failures. *The Economist* asserts that AI was once "associated with hubris and disappointment" but it has "suddenly become the hottest field in technology" (Standage, 2016, p. 1). However, AI research and development efforts have often been construed by many business analysts in ways that neglect entirely to mention the failures and wastes of resources associated with their pasts (Ramiller, 2006).

Recent discourses concerning AI and automation have strong linkages to that of the 1950s and 1960s in their emphases on technological change and job loss as well as on the potential ascendancy of thinking machines:

> History can shed some light on our concerns. It was in the middle of the last century that the United States last seemed to encounter job destroying technologies on today's scale. (The economic woes of the 1970s and '80s were mostly blamed--at least in the popular mind-on Japanese imports.) Automation was a hot topic in the media and among social scientists, pundits, and policymakers. It was a time of unsettlingly rapid technological change, much like our own. (Akst, 2013, p. 68)

Some social science research has shown that survey subjects are expressing "us vs. them" attitudes in which AI is construed as a foreign and somewhat menacing "other" (Jepsen et al., 2016; Oh et al., 2017).

This guarded tone has emerged in discussions of some specific recent installations of AI technologies as well. In their analysis of the operations of DeepMind (a British-based AI subsidiary of Google/Alphabet), Powles and Hodson (2017) state that "digital pioneers who claim to be committed to the public interest must do better than to pursue secretive deals and specious claims in something as important as the health of populations" (p. 14). Kaplan (2016) directly critiques the "gee whiz" nature of many accounts of AI research and applications (p. 37).

Many of the current applications for AI have the potential to alter employment opportunities, involving unsettling social transformations, for example as presented in Hakken et al. (2015), Wajcman (2017), and McChesney and Nichols (2016). The McChesney and Nichols book counsels that "people get ready" for devastating societal changes (p. 1). In recent years, the notion of "sociopolitical cybernetics" (Armitage, 2016; Virilio, 2010) has emerged to characterize the potential erosion of control held by many individuals of the growth and dissemination of societal practices associated with many AI systems. Some AI applications, framed in terms of "bots," have played increasingly salient roles in social media and financial contexts with little human oversight (Wooley, 2018), leading to potential disruption of critically-important social and political processes. The issues involved may have broad ramifications but also require well-grounded and specific analyses: for instance, Geist (2016) expresses his concerns about the "AI arms race" in which autonomous systems will make critical judgments about the deployment of nuclear weapons. Taddeo and Floridi (2016) contend that the moral responsibilities of those involved with intelligent systems and networks are linked to these individuals' "gatekeeping function, their corporate social responsibilities, and their role in implementing and fostering human rights" (p. 1575).

Examples of Success- and Failure-Related AI Narratives

AI attracted considerable academic, corporate, and military attention in the late 1950s and early 1960s as prototypes and pilot projects emerged within a short period of time. However, many of the early demonstration projects of AI in these decades soon failed to live up to their initial promise; some researchers subsequently worked actively to distance themselves from the apparently "outrageous" claims for the future of AI, preferring to frame their own efforts in more constrained ways (Stafford, 2001). Many of the founding AI strategies focused on the capture of the expertise of individuals in "knowledge engineering," with some literally labeled "experts in a box" (Oravec, 2014; Oravec & Travis, 1992). The hyperbolic assertions that AI systems would soon displace many human experts such as medical doctors were often seen as problematic to the professionals who were seeking appropriate applications for technology rather than the production of "black boxes" in which something labeled as "knowledge" would be stored (Perrolle, 1991). Discourse in academic and popular culture arenas in the 1970s and 1980s often focused on whether AI systems could indeed "think" as well as on the prospects for the replacement of human experts with intelligent machines. Barr and Feigenbaum (1982) stated that an AI system "should be able to learn what human experts know, so that it can perform as well as they do, understand the points of departure among the views of human experts who disagree, keep its knowledge up to date as human experts do (by reading, asking questions, and learning from experience), and present its reasoning to its human users in much the way that human experts would (justifying, clarifying, explaining and even tutoring)" (p. 80). Many of the AI efforts in the 1970s and 1980s also had strong and direct ties to Tayloristic (scientific management) perspectives and the routinization of human expertise (Kraft, 1997; Stafford, 2001). Whether or not these Tayloristic efforts were successful in changing workplaces, they often affected how individuals characterized their own vocations and planned their occupational futures, including in some information-oriented professions (Banja, 2020).

As previously outlined, many negative and sobering themes involving AI history and applications have been less well covered in its current academic and marketing-related characterizations, although they may be useful in sense-making about technologies (Fincham, 2002). The UK's Lighthill Report of 1974 provided a devastating attack on artificial intelligence that directly resulted in the nearly "complete dismantling" of AI research in England (Crevier, 1993); exaggerated claims made for AI were reportedly a part of this backlash. AI as a research and development arena often faced "AI winters," periods in which little attention was paid to it and research funding became scarce. In efforts to avoid AI winters, commercialization of AI became more of an imperative: "With the expert system boom in full swing, epistemology was brought into public view and was shown to have commercial value" (Stafford, 2001).

The "AI business" of the early 1980s (characterized in Winston & Prendergast, 1984) drew many researchers away from basic research and into start-up organizations or corporate settings. Criticisms of the apparent hyperbole associated with AI efforts during this period abound. Davenport and Prusak (1998) state that "The field of knowledge technology has suffered from overly high expectations and excessive levels of hype, particularly with regard to expert systems" (p. 126); they note the "limited success" of these early efforts, which Blair (2002) also characterizes in terms of "hype" and a "gold rush." The hype involved with AI often overshadowed the positive gains AI researchers were able to achieve through rigorous and sustained research efforts.

One of the early exemplars of an AI implementation that produced high levels of publicity and hyperbole but little long-term gain was developed for the US Campbell Soup Company in 1985 by Dr. Richard Herrod of Texas Instruments. The system was designed to gauge how long to cook specific varieties of soup (Mans, 1995). One of the experts assigned to this effort (Aldo Cimino) was retiring after forty-four years, so the system was reportedly designed to replace him rather than training humans to do his duties. Below is a description of this initiative in a trade publication:

> Campbell's maintenance person, Aldo Cimino, had 44 years of experience on the giant hydrostatic cookers that sterilize soup and other canned products. He knew more than anyone in the company about these complex pieces of equipment and was called in to consult at plants around the world. (Mans, 1995, p. 16)

Specific rationales for eliciting Cimino's insights through technological means rather than training interns or apprentices are unclear. However, the following account provides a narrative of Campbell's AI approach:

> One day his boss woke up and realized that Aldo was going to retire soon, and then who would answer the tough questions? He called in Dr. Herrod and his group to capture Aldo's expertise. The project took six months and endless man-hours, but resulted in an expert system capable of answering most of the problems as well as Aldo could. (Mans, 1995, p. 16)

The Campbell Soup initiative was terminated before it became profitable to the corporation. Efforts to use AI systems in everyday sales efforts extended into the 1990s at Famous Footwear in the US but were discontinued as well ("Hype is Gone," 1995).

An assortment of influential books and conference presentations that were highly critical of AI developers levied specific attacks on their associated hyperbolic assertions in the latter part of the twentieth century; unfortunately, these items are less often cited today than their more positive and enthusiastic counterparts. They include *Man over Machine: The Power of Human Intuition and Expertise in the Era of the Computer* (Dreyfus & Dreyfus, 1986) and *Computer Power and Human Reason: From Judgment to Calculation* (Weizenbaum, 1976), following earlier, pioneering works such as Taube's (1963) *Computers and Common Sense*. Authors of these books often had considerable technological background in AI, with Joseph Weizenbaum a noted professor of AI at the Massachusetts Institute of Technology (MIT) and Hubert Dreyfus a philosophy professor at Stanford with expertise in cognitive science. They attempted to counter what they considered the more outlandish claims of AI proponents as well as efforts to integrate AI into applications in ways that they claimed inappropriately substituted AI-enhanced systems for human labor. Insights from the various failures related to AI (such as the ones described in this section and in the above-mentioned books) could be more useful in unpacking critical concerns than many of today's overtly technology-apologetic narratives. As an example of the latter, some recent commentators have outlined a "moral imperative" for the application of AI in everyday contexts (Vardi, 2016), and advise workers on *What To Do When Machines Do Everything: How to Get Ahead in a World of AI, Algorithms, Bots, and Big Data* (Frank et al., 2017). Still others have characterized what is reportedly necessary to be "robot-proof" in higher education contexts, with a book of that title (Aoun, 2017).

Expanding the Dimensions of Corporate Responsibility for AI: A "Robot Tax"?

As related in this chapter, concerns about technological impacts on society and on job losses in particular have a long history: discourses about the relative places of humans and machines in economic and social realms have also emerged in strong forms recently (McChesney & Nichols, 2016; Wajcman, 2017). However, for many years, there was little engagement with critically-important employment and social welfare issues by the AI researchers and developers at Google/Alphabet, Facebook, Amazon,

IBM, and Apple who actively designed and implemented AI technologies (Oravec, 2018, 2019). Fortunately, this neglect of social concerns has changed somewhat. For example, there was an upsurge in corporate interest in the ethics of AI in late 2016 with the formation of these corporations' "Partnership for AI" (Richards, 2017), reportedly a response to increased public attention to this arena. A number of corporate and governmental efforts to understand robotic and AI implications for society have followed in the past several years (as related in the last chapter of this book). Through the decades, however, the corporations involved benefited from the uncertainty and confusion associated with technological disruption (Powles & Hodson, 2017). For example, by framing many debates about technology-related ethics with the motto "don't be evil" as a focus, Google shifted the discourse to larger-grained concerns and perhaps even theology (Oravec, 2018), often avoiding more delicate and nuanced moral analyses about human wellbeing such as those provided by Ozer (2012). Many important issues concerning AI and related technologies have been framed by some corporations as epideictic and rhetorical rather than dealing with specific and critical human rights and welfare challenges such as job deskilling, privacy, and AI-reinforced bias (Ekbia & Nardi, 2017; Fahnestock, 1986).

Forms of discourse rooted in simplistic mottos and technological hyperbole have often framed Google/Alphabet and other high-tech organizations' ethical situations in terms of exceptionalism, rather than being rooted in the traditions of moral thinking in technology-related ethics. These organizations may have behaved differently if they had generated, disseminated, and championed detailed corporate ethics codes rather than the "don't be evil" motto and comparable expressions as well as funded research in the societal issues associated with AI and automation. For example, few high-tech organizations have explored the notion that the personal data and many interactions contributed by individuals to AI systems form a sort of community investment in these technologies that deserves either direct financial return or at least some significant stake in decision making concerning the technologies (Cheney-Lippold, 2017). The investment that governments have made in basic and applied research in robots, autonomous vehicles, and AI (funded by taxpayers) should also reap some reward for the public sphere. As society moves toward public policies that incorporate "abundance economics" notions, ways of conscientiously allocating the gains from technologies across societal levels are being deliberated (Swan, 2017). The notion of a "robot tax," proposed by Mady Delvaux-Stehres and endorsed by Microsoft founder Bill Gates, has recently catalyzed interest in the prospects for redistribution of some AI-related profits into education and social welfare (Rimmer, 2017). The notion has been opposed, however, as diminishing the intrinsic value of work (Moser, 2021). The proposition that great public investment went into AI (in the form of tax dollars as well as data input) and should be matched with some level of public compensation could be empowering to those who wish to moderate the apparent avarice and societal neglect of many commercial technology profiteers.

Conclusions and Reflections

The heady or even dizzying appeal of AI hype may indeed influence many people to adopt AI-related products and services and perhaps overlook some of their foibles. However, efforts to moderate AI characterizations and rhetorical approaches so that hyperbole is replaced by more specific and nuanced analyses are helpful in resolving many critical issues involved with robotic- and AI-related impacts. Although hyperbole and "dazzle" (Lui, 2018) are often needed to attract corporate funding and academic research grants, straightforward and commonsense approaches may have more lasting value in the long run both for organizations' and individuals' wellbeing. This is not to say that technological vision is unimportant, nor that the motivation of researchers and developers should be neglected; futuristic dreams of homes, workplaces, and communities that are empowered by capable and appropriate robotic and AI assistance should certainly not be inhibited. However, such hyperbole-ridden efforts as developers' designing a robot or AI system to win a "contest" against an individual, or conducting demonstrations to show how AI-enhanced entities can "outclass" humans, are far too common. Various game shows, challenges, and well-publicized chess matches have formed a good deal of the rhetorical framing of robotic and AI futures, rather than demonstrations of human support and nurturance. Unfortunately, recent technological successes in some of these competitions have buttressed the claims of developers that AI applications and algorithms should indeed be widely implemented, despite social costs, implementation complications, and moral concerns. However, the projections and speculations about AI can indeed be driven in measured ways by broader societal concerns and not primarily by individual or corporate-level opportunism. A number of technological apologists have provided characterizations of AI that do little to elucidate critical aspects of the technologies and enable individuals to make well-considered decisions about their uses. As such applications as self-driving cars and autonomous security systems become more common, realistic and nuanced characterizations of AI are required in order to make sound decisions about safety and environmental impact in our everyday lives.

Social and historical perspectives on AI from previous generations of researchers and analysts have played roles in recent ethical and social analyses, often embellishing AI overstatement with injections of reflections from prior eras. The era in which AI approaches emerged as an influence in research and development (from the 1950s to the early 1970s) produced an assortment of grand technological projections that served to shape deliberation on business strategies, military research directions, and public policy agendas, declarations that are often injected in AI-related discourse (Armstrong et al.2014; Oravec, 2019). Artificial intelligence research has indeed had some considerable successes in past decades, as predicted by its progenitors: for example, chess-playing computers can defeat grand masters, as projected by a number of AI pioneers (Arnold & Scheutz, 2016). The far less palatable message delivered in many accounts of how AI will affect society today involves the systematic devaluation of human effort, something that could have unfortunate cultural side effects as well as the direct implications of large-scale unemployment. For example, Barrat (2013) characterizes AI as "our final invention" as a species and laments "the end of the human era" (p. 1).

As discussed in this chapter, the "thinking machine" notion served to energize a good deal of early AI and robotics research initiatives and has continued as a theme in many current attempts to package AI notions for various audiences. Ada Lovelace, often construed as the "first programmer," projected early thinking machine notions in the 1800s (Morais, 2013). Today's physical sensors and social media advances have injected more data and increased processing power into "intelligent" systems, which has advanced the prospects for AI to be infused into everyday workplace and community processes (Makridakis, 2017). These enhanced thinking machines have emerged, at least in some recent technological imaginaries, with sufficient contest and gaming "wins" to reinforce their supposed superiority in particular intellectual realms, such as trivia contests and chess (Bucher, 2017). However, extension of victories in such restrictive contest and game settings into the license for indiscriminate proliferation of technologies is problematic. Historical legacies (such as those provided in this chapter) may seem to be irrelevant to fast-moving technologies; rather, reflection on both the successes and failures involving critical systems can have implications in the teaching of students as well as research (Banks, 2016).

> The world of the future will be an even more demanding struggle against the limitations of our intelligence, not a comfortable hammock in which we can lie down to be waited upon by our robot slaves. (Norbert Wiener, 1966)

In the context of AI, students' training may put them into roles in which they will have influence on AI systems' future directions; they may soon have the opportunity to participate in active and embedded ethical work in this arena (Goldsmith & Burton, 2017) and can work to stem the flow of AI hype. Discourse about the hyperbolic dimensions of AI and robotics generated through the decades by researchers, developers, community participants, and public policy leaders holds insight for current decision makers as well as signals some potentials for future discord.

References

Akst, D. (2013). Automation anxiety. *The Wilson Quarterly (1976)*, *37*(3), 66–78.

Aoun, J. E. (2017). *Robot-proof: Higher education in the age of artificial intelligence*. MIT Press.

Armitage, J. (2016). Vision, inertia, and the mobile telephone: On the origins of control space and the spread of sociopolitical cybernetics. In J. Macgregor Wise & H. Koskela (Eds.), *New visualities, new technologies: The new ecstasy of communication*, pp. 67–82. Ashgate.

Armstrong, S., Sotala, K., & Ó hÉigeartaigh, S. S. (2014). The errors, insights and lessons of famous AI predictions–and what they mean for the future. *Journal of Experimental & Theoretical Artificial Intelligence*, *26*(3), 317–342.https://doi.org/10.1080/0952813X.2014.895105.

Arnold, T., & Scheutz, M. (2016). Against the moral turing test: Accountable design and the moral reasoning of autonomous systems. *Ethics and Information Technology*, *18*(2), 103–115.

Banja, J. (2020). AI hype and radiology: A plea for realism and accuracy. *Radiology: Artificial Intelligence*, *2*(4), e190223. https://doi.org/10.1148/ryai.2020190223

Banks, S. (2016). Everyday ethics in professional life: Social work as ethics work. *Ethics and Social Welfare*, *10*(1), 35–52.

Barr, A., & Feigenbaum, E. A. (1982). *Handbook of artificial intelligence*, Vol. 2. Addison Wesley.
Barrat, J. (2013). *Our final invention: Artificial intelligence and the end of the human era*. Macmillan.
Blair, D. C. (2002). Knowledge management: Hype, hope, or help? *Journal of the American Society for Information Science & Technology, 53*(12), 1019–1028.
Bolisani, E. (2008). Knowledge transfer on the net: Useful lessons from the knowledge economy. In E. Bolisani (Ed.), *Building the knowledge society on the internet: Sharing and exchanging knowledge in networked environments*, pp. 110–128. IGI Global. https://doi.org/10.4018/978-1-59904-816-1
Bollier, D. (2017). *Artificial intelligence comes of age*. Aspen Institute. https://www.aspeninstitute.org/publications/artificial-intelligence-comes-age/
Bruderer, H. (2017). Computing history beyond the UK and US: Selected landmarks from continental Europe. *Communications of the ACM, 60*(2), 76–84.
Brynjolfsson, E., & McAfee, A. (2011). *Race against the machine*. Digital Frontier.
Bucher, T. (2017). The algorithmic imaginary: Exploring the ordinary effects of Facebook algorithms. *Information, Communication & Society, 20*(1), 30–44.
Caulfield, T. (2000). Underwhelmed: Hyperbole, regulatory policy, and the genetic revolution. *McGill Law Journal, 45*(2), 437–460.
Cheney-Lippold, J. (2017). *We are data: Algorithms and the making of our digital selves*. NYU Press.
Cohen, H., Cohen, B., & Nii, P. (1983). *The first artificial intelligence coloring book*. Addison-Wesley Publishers.
Coleman, M. (2016). Paralogical hyperbole: A "missing link" between technical and public spheres. *Poroi, 12*(1), 1–24.
Conway, F., & Siegelman, J. (2004). *Dark hero of the Information Age: In search of Norbert Wiener the father of Cybernetics*. Basic Books.
Cordeschi, R. (2007). AI turns fifty: Revisiting its origins. *Applied Artificial Intelligence, 21*(4–5), 259–279.
Crevier, D. (1993). *AI: The tumultuous search for artificial intelligence*. BasicBooks.
Davenport, T. H., & Prusak. L. (1998). *Working knowledge: How organizations manage what they know*. Harvard Business School Press.
Dreyfus, H., & Dreyfus, S. (1986). *Man over machine: The power of human intuition and expertise in the Era of the Computer*. Macmillan.
Ekbia, H., & Nardi, B. (2017). *Heteromation, and other stories of computing and capitalism*. MIT Press.
Fahnestock, J. (1986). Accommodating science: The rhetorical life of scientific facts. *Written Communication, 3*(3), 275–296. https://doi.org/10.1177/0741088386003003001

Feigenbaum, E. A., & McCorduck, P. (1984). *The fifth generation*. Signet.
Fincham, R. (2002). Narratives of success and failure in systems development. *British Journal of Management, 13*(1), 1–14.
Forsythe, D. (2001). *Studying those who study us: An anthropologist in the world of artificial intelligence*. Stanford University Press.
Frank, M., Pring, B., & Roehrig, P. (2017). *What to do when machines do everything: How to get ahead in a world of AI, algorithms, bots, and big data*. John Wiley & Sons.
Geist, E. M. (2016). It's already too late to stop the AI arms race—We must manage it instead. *Bulletin of the Atomic Scientists, 72*(5), 318–321.
Goldsmith, J., & Burton, E. (2017, February). Why teaching ethics to AI practitioners is important. In *AAAI Proceedings 2017*, pp. 4836–4840. https://aaai.org/ocs/index.php/AAAI/AAAI17/paper/view/14271
Hakken, D., Teli, M., & Andrews, B. (2015). *Beyond capital: Values, commons, computing, and the search for a viable future*. Routledge.
Helbing, D., Frey, B. S., Gigerenzer, G., Hafen, E., Hagner, M., Hofstetter, Y., van den Hoven, J., Zicari, R., & Zwitter, A. (2017, February 25). Will democracy survive big data and artificial intelligence? *Scientific American*. http://hdl.handle.net/11858/00-001M-0000-002C-F151-2
Heilbroner, R. L. (1965). Men and machines in perspective. *The Public Interest, 1*, 27.
Hellström, T., & Raman, S. (2001). The commodification of knowledge about knowledge: Knowledge management and the reification of epistemology. *Social Epistemology, 15*(3), 139–154.
Heylighen, F., & Lenartowicz, M. (2017). The Global Brain as a model of the future information society. *Technological Forecasting and Social Change, 114*, 1–6. https://doi.org/10.1016/j.techfore.2016.10.063
Hollnagel, E. (2014). Human factors/ergonomics as a systems discipline? *The Human Use of Human Beings* revisited. *Applied Ergonomics, 45*(1), 40–44.
Hong, S. (2004). Man and machine in the 1960s. *Techné: Research in Philosophy and Technology, 7*(3), 50–78.
Hughes, C., & Hughes, T. (2019). What metrics should we use to measure commercial AI? *AI Matters, 5*(2), 41–45.
Hurlburt, G. (2017). Superintelligence: Myth or pressing reality? *IT Professional, 19*(1), 6–11.
"Hype is gone, but job of expert systems remains." (1995). *Chain Store Age, 71*(10), 66.
Jepsen, K. S., Delgado, A., & Bertilsson, T. (2016). 'The public spectre': A critical concept of public engagement with technology. In *Technoscience and Citizenship: Ethics and Governance in the Digital Society*, pp. 17–30. Springer International Publishing.

Johnson, D. G., & Verdicchio, M. (2017). Reframing AI discourse. *Minds and Machines, 27*(4), 575–590.
Johnston, J. (2008). *The allure of machinic life: Cybernetics, artificial life, and the new AI.* MIT Press.
Kaplan, J. (2016). Artificial intelligence: Think again. *Communications of the ACM, 60*(1), 36–38.
Kline, R. R. (2015). *The cybernetics moment: Or why we call our age the information age.* JHU Press.
Kraft, P. (1997). *Programmers and managers: The routinization of computer programming in the United States.* Springer-Verlag.
Kling, R. (1991). Computerization and social transformation. *Science, Technology and Human Values, 17,* 342–367.
Langlois, R. N. (2003). Cognitive comparative advantage and the organization of work: Lessons from Herbert Simon's vision of the future. *Journal of Economic Psychology, 24*(2), 167–187.
Lee, K. Y., & Kim, J. (2016). Artificial intelligence technology trends and IBM Watson references in the medical field. *Korean Medical Education Review, 18*(2), 51–57.
Lehman-Wilzig, S. N. (1981). Frankenstein unbound: Towards a legal definition of artificial intelligence. *Futures, 13*(6), 442–457.
Liu, H. Y. (2018). The power structure of artificial intelligence. *Law, Innovation and Technology, 10*(2), 197–229.
McCarthy, J., Minsky, M. L., Rochester, N. & Shannon, C. E. (1955, August 31). A proposal for the Dartmouth summer research project on artificial intelligence. http://www-formal.stanford.edu/jmc/history/dartmouth/dartmouth.html
McChesney, R., & Nichols, J. (2016). *People get ready: The fight against a jobless economy and a citizenless democracy.* Nation Books.
Makridakis, S. (2017). The forthcoming artificial intelligence (AI) revolution: Its impact on society and firms. *Futures, 90,* 46–60.
Mans, J. (1995, April). The new experts. *Dairy Foods,* 15–20.
Markoff, J. (2013, May 21). In 1949, He imagined an age of robots. *New York Times,* D8.
Metz, C. (2018, February 12). How artificial intelligence is edging its way into our lives. https://www.nytimes.com/2018/02/12/technology/artificial-intelligence-new-work-summit.html
Miller, A. (1992). Check and mate: The ancient game meets modern times. *Compute! 14*(9), 91.
Morais, B. (2013, October 15). Ada Lovelace, the first tech visionary. *New Yorker.* https://www.newyorker.com/tech/annals-of-technology/ada-lovelace-the-first-tech-visionary

Morris, K. C., Schlenoff, C., & Srinivasan, V. (2017). A remarkable resurgence of artificial intelligence and its impact on automation and autonomy. *IEEE Transactions on Automation Science and Engineering, 14*, 407–409.

Moser, E. (2021). Against robot taxes: Scrutinizing the moral reasons for the preservation of work. *AI and Ethics, 1*(4), 491–499.

Nemesi, A. L. (2004). What discourse goals can be accomplished by the use of hyperbole? *Acta Linguistica Hungarica, 51*(3–4), 351–378.

Newlands, G. (2021). Lifting the curtain: Strategic visibility of human labour in AI-as-a-Service. *Big Data & Society, 8*(1). https://doi.org/10.1177/20539517211016026

Noble, D. F. (1995). *Progress without people: New technology, unemployment, and the message of resistance*. Between the Lines.

Oh, C., Lee, T., Kim, Y., Park, S., Kwon, S., & Suh, B. (2017). Us vs. them: Understanding artificial intelligence technophobia over the Google DeepMind Challenge Match. In *Proceedings of the 2017 CHI Conference on Human Factors in Computing Systems*, 2523–2534. ACM Press.

Olson, H. (2001). Patriarchal structures of subject access and subversive techniques for change. *Canadian Journal of Information & Library Sciences, 26*(2/3), 1–29.

Oravec, J. A. (2014). Expert systems and knowledge-based engineering (1984–1991): Implications for instructional systems research. *International Journal of Designs for Learning, 5*(2), 65–74.

Oravec, J. A. (2018). "Don't be evil" and beyond for high tech organizations: Ethical statements and mottos (and responsibility). In S. Chhabra (Ed.), *Civic Engagement and Social Change in Contemporary Society*. IGI Global.

Oravec, J. A. (2019). Artificial intelligence, automation, and social welfare: Some ethical and historical perspectives on technological overstatement and hyperbole. *Ethics and Social Welfare, 13*(1), 18–32.

Oravec, J. A., & Travis, L. (1992). If we could do it over, we'd ... Learning from less-than-successful expert system projects. *Journal of Systems and Software, 19*(2), 113–122.

Ostergren, J. E., Dingel, M., McCormick, J., & Koenig, B. (2015). Unwarranted optimism in media portrayals of genetic research on addiction overshadows critical ethical and social concerns. *Journal of Health Communication, 20*(5), 555–565.

Ozer, N. (2012). Putting online privacy above the fold: Building a social movement and creating corporate change. *New York University Review of Law & Social Change, 36*, 215–281.

Pankewitz, C. (2017). Automation, robots, and algorithms will drive the next stage of digital disruption. In A. Khare, B. Stewart, & R. Schatz (Eds.), *Phantom ex machina* (pp. 185–196). Springer International Publishing.

Perrolle, J. (1991). Expert enhancement and replacement in computerized mental labor. *Science, Technology, & Human Values, 16*(2), 195–207.
Porat, M. (1977). *The information economy.* US Department of Commerce.
Powles, J., & Hodson, H. (2017). Google DeepMind and healthcare in an age of algorithms. *Health and Technology, 7*(4), 351–367.
Ramiller, N. C. (2006). Hype! Toward a theory of exaggeration in information technology innovation. In *Academy of Management Proceedings,* (1), A1-A6. Academy of Management.
Ranchordás, S. (2017). Digital agoras: Democratic legitimacy, online participation and the case of Uber-petitions. *The Theory and Practice of Legislation, 5*(1), 31–54.
Richards, J. (2017). The cognitive era. *ITnow, 59*(2), 8–9.
Richardson, K. (2015). *An anthropology of robots and AI: Annihilation anxiety and machines.* Routledge.
Rifkin, J. (1995). *The end of work: Technology, jobs, and our future.* Putnam.
Rimmer, M. (2017). The Wild West of Robot Law. *Australasian Science, 38*(3), 20–21.
Rosenblatt, F. (1958). The perceptron: A probabilistic model for information storage and organization in the brain. *Psychological Review, 65*(6), 386–408.
Saygin, A. P., Cicekli, I., & Akman, V. (2000). Turing Test: 50 years later. *Minds and Machines, 10*(4), 463–518.
Simon, H. A. (1995). Artificial intelligence: An empirical science. *Artificial Intelligence, 77*(1), 95–127.
Simon, H. A. (1996/1961). *The Sciences of the Artificial.* MIT Press.
Simon, H. A., & Newell, A. (1958). Heuristic problem solving: The next advance in operations research. *Operations Research, 6*(1), 1–10.
Simpson, B. (2016). Algorithms or advocacy: Does the legal profession have a future in a digital world? *Information & Communications Technology Law, 25*(1), 50–61.
Stafford, S. P. (2001). Epistemology for sale. *Social Epistemology, 15*(3), 215–230. https://doi.org/10.1080/02691720110076549
Standage, T. (2016, June 25). Special report, artificial intelligence, the return of the machinery question. *The Economist,* 1–5.
Swan, M. (2017). Is technological unemployment real? An assessment and a plea for abundance economics. In I. K. LaGrandeur & J. J. Hughes (Eds.), *Surviving the machine age* (pp. 19–33). Springer International Publishing.
Szathmári, I. (1958). The stylistic investigation of word meaning. In F. Terestyéni (Ed.), *An Outline of Hungarian Stylistics* (pp. 65–148). Tankönyvkiadó.
Taddeo, M., & Floridi, L. (2016). The debate on the moral responsibilities of online service providers. *Science and Engineering Ethics, 22*(6), 1575–1603.
Taube, M. (1963). *Computers and common sense.* McGraw-Hill.

Turing, A. M. (1950). Computing machinery and intelligence. *Mind, LIX, 236,* 433–460.

Ulnicane, I., Knight, W., Leach, T., Stahl, B. C., & Wanjiku, W. G. (2021). Framing governance for a contested emerging technology: Insights from AI policy. *Policy and Society, 40*(2), 158–177. https://doi.org/10.1080/14494035.2020.1855800

van der Laan, J. M. (2016). The dominant narrative. In J. M. van der Laan (Ed.), *Narratives of Technology* (pp. 41–73). Palgrave Macmillan.

Van Nostrand, A. D. (2013). *Fundable knowledge: The marketing of defense technology.* Routledge.

Vardi, M. Y. (2016). The moral imperative of artificial intelligence. *Communications of the ACM, 59*(5), 5.

Virilio, P. (2010). *The futurism of the instant: Stop-eject.* Polity.

Vogel, L. (2017). Plan needed to capitalize on robots: AI in health care. *Canadian Medical Association Journal, 189*(8), E329–E330. https://doi.org/10.1503/cmaj.1095395

Wajcman, J. (2017). Automation: Is it really different this time? *British Journal of Sociology, 68*(1), 119–127.

Weizenbaum, J. (1976). *Computer power and human reason: From judgment to calculation.* WH Freeman.

Wiener, N. (1960). Some moral and technical consequences of automation. *Science, 131*(3410), 1355–1358.

Wiener, N. (1966). *God & Golem, Inc.: A comment on certain points where cybernetics impinges on religion.* MIT press.

Wiener, N. (1988/1950). *The human use of human beings: Cybernetics and society* (No. 320). Perseus Books Group.

Winston, P., & Prendergast, K. A. (1984). *The AI business. The commercial uses of artificial intelligence.* MIT Press.

Woolgar, S. (1989). Reconstructing man and machine: A note on sociological critiques of cognitivism. In W. E. Bijker, T. P. Hughes, & T. Pinch (Eds.), *The Social Construction of Technological Systems* (pp. 311–328). MIT Press.

Woolley, S. (2018). The political economy of bots: Theory and method in the study of social automation. In *The Political Economy of Robots,* pp. 127–155. Palgrave Macmillan.

CHAPTER 7

"Our Hearts Go Out to the Victim's Family": Death by Robot and Autonomous Vehicle

Robots and other AI-enhanced, autonomous entities have indeed been a part of an assortment of social and economic advances over the past decades, as shown in numerous technical and financial reports (Abnet, 2020; George, 2016). However, there have been considerable human costs for these initiatives, along with incidents and phenomena that are clearly on the "dark side." The following unsettling paragraph can be found in 2017 US District Court documents:

> Upon information and belief, Wanda [Holbrook] was working in either section 140 or 150 within the '100' cell, when a robot from section 130 took Wanda by surprise, entering the section she was working in... Upon entering the section, the robot hit and crushed Wanda's head between a hitch assembly it was attempting to place in the fixture of section 140, and a hitch assembly that was already in the fixture, and Wanda suffered tremendous fright, shock and conscious pain.... (*Courthouse News*, 2017)

Despite the devastating robot-inflicted injuries that led to Wanda Holbrook's death (and potential corporate liability), the company at which Wanda Holbrook worked was fined only seven thousand US dollars by State of Michigan regulators (Associated Press, 2016).

Deaths and injuries such as Wanda Holbrook's are often, unfortunately, among the grim side effects of production and mobility. These deaths and injuries have not ceased to be a factor despite decades of efforts on

© The Author(s), under exclusive license to Springer Nature Switzerland AG 2022
J. A. Oravec, *Good Robot, Bad Robot*, Social and Cultural Studies of Robots and AI, https://doi.org/10.1007/978-3-031-14013-6_7

safety and risk management. The expression "Our hearts go out to the victim's family" is an all-too-common utterance of organizational representatives. For example, it was attributed to Uber spokeswoman, Sarah Abboud after a death caused by an autonomous vehicle (as reported in Wakabayashi, 2018). Deaths and injuries that are associated with robots and other autonomous entities are often placed in a different light than other sorts of incidents in dramaturgical perspective; the sense of these deaths as being engendered outside of human control often intensifies their personal and social impacts. Themes and images of murder, domination, and malice introduced from science fiction and popular discourse often emerge along with expressed feelings of creepiness that are akin to those associated with monsters and zombies (Watt et al., 2017). Trauma associated with these robotic attacks for onlookers, associates, and first responders can be devastating and have lasting impacts. Potentials for associated anti-robot backlash and security breaches have apparently also increased, despite extensive research on how to make robots more palatable and attractive to human workers. Since robots, autonomous vehicles, and other AI-enhanced entities are promoted as saving lives, these deaths take on special significance. According to Meta's CEO Mark Zuckerberg, "If you recognize that self-driving cars are going to prevent car accidents, AI will be responsible for reducing one of the leading causes of death in the world" (quoted in Sharma, 2021).

This chapter explores how deaths and injuries by robots and autonomous systems have been distinguished from other kinds of lethal incidents; it examines the implications of these assignments for how the incidents are handled in terms of safety and risk assessment, as well as in discourse on work itself. The kinds of methodical and detailed after-crash analyses applied to airline accidents in the US and some other nations are needed to analyze incidents related to autonomous entities, though the need to access proprietary information and involve multiple outsources can often stymie investigatory efforts. The deep fear and distrust that is often coupled with these lethal incidents by those involved either directly or as onlookers is a part of the robot and AI "overload" that can have a severe psychological toll. Consider the psychological impact of the following account:

> The New York Police Department said that early on July 26, a 52-year-old man was hit by a Tesla and killed while changing a flat tire on his vehicle, which was parked on the left shoulder of the Long Island Expressway in

Queens. The Tesla driver remained at the scene, and the department's Collision Investigation Squad is working on the case, the release said. The department identified the victim as Jean Louis of Cambria Heights, New York. (CNBC, 2021)

As the varieties and numbers of robots and autonomous systems (including individuals' prostheses) increases, variations in the narrative themes associated with these deaths are developing and initiatives to foster more accepting attitudes on the part of humans to robots have expanded, despite the value of the human survival tendencies that are linked with appropriate distrust and distance. Robots and AI are also increasingly being integrated into the processes involved with human death and burial (Gould et al., 2021), further strengthening the linkages between robotics and human mortality. Some efforts to memorialize robot deaths with funerals (especially in the military) are emerging as well (Garber, 2013).

The "death by robot" narrative theme is powerful: it is often employed in efforts to characterize workplace and infrastructure automation issues, including the prospects for subsequent anti-robot sabotage or destruction on the part of workers. Promotional efforts (including initiatives that are designed to make robots appear friendly and humanlike) rather than protective design strategies and related social discourse could unfortunately influence workers to be at ease with robots that are potentially unpredictable and dangerous. For example, as service and delivery robots speed through safe-appearing corridors such as sidewalks and hallways, the prospects for injury and even death increase (Salvini et al., 2021). The robot-AI-human death connection has the potential to shape the character of many workplaces and communities, but there may be a future in which the "wildness" and eccentricities of both robots and humans can coexist safely and be respected. These issues also affect how autonomous vehicles are being integrated into society; the growing numbers and kinds of self-driving car deaths are creating legal and social concerns for automobile corporations (Wakabayashi, 2018). In 2018, Elaine Herzberg was bicycling in Tempe, Arizona when struck by an Uber (Cellan-Jones, 2020). Her death is often considered the first death by autonomous vehicle in the US, and many have followed. The fact that bicyclers, pedestrians, first responders, and individuals in cars pulled over on the side of the road are often the ones who face the greatest peril through self-driving vehicles underscores the potential for emerging "death disparities"

(Nunes & Axhausen, 2021), inequities that should be examined carefully in allocations of research dollars in the arena,

A Sad Compendium of Deaths by Robots and Autonomous Entities

On January 25, 1979, Robert Williams reportedly died as a result of interaction with a robot. He was employed at a Ford Motor Company casting plant (Winfield et al., 2020). On July 4, 1981, Kenji Urada, a maintenance worker at the Kawasaki Heavy Industries plant in Akashi, Japan, was killed while in the process of dealing with a robot malfunction (UPI, 1981). On June 19, 2016, a worker in a metal stamping plant, Regina Elsea, died of robot-inflected injuries suffered the day before (Berry, 2017). Many other workplace deaths have followed with connections to robots and autonomous systems, with a variety of reactions and analyses (Lyons, 2018; Ssanderson et al., 1986). For example, in India, a 23-year-old factory employee Ramji Lal was crushed to death by industrial robot in an auto component manufacturing factory (Sahni, 2015). One of the reactions to his death is the following: "It is a case of negligence by the company. In a situation where robots and heavy machinery is installed, there should be proper sensors so that the robot stops when a person comes too close to it" (Kuldeep Janghu, general secretary of Maruti Udyog Kamgar Union, as cited in a *Hindustan Times* article). In turn, various attacks upon robots by humans have also been conducted (Bartneck & Keijsers, 2020), as discussed in the next chapter. These interactions have a long history that extends from fiction and film to unfortunate real-life human demise. Later in this book, deaths by such government-authorized entities as police robots and military drones will be explored in depth, incidents that open comparably challenging social and moral issues. This chapter analyzes human responses to these workplace and community incidents as well as some potentially-problematic research and development efforts designed to make robots less fearful for humans. It characterizes an assortment of workplace deaths that are the result of contact with autonomous entities in hopes of managing change in workplaces and communities while mitigating the dangers associated with automation.

The connections explored in the sections to come have a great deal to do with the evolving character of everyday settings as automation plays larger roles in transportation, manufacturing, and service operations. Many workplace and community sites are being redesigned so that they are better fit for robotic and automated vehicle utilization, with fewer accommodations for human activity. The casual familiarity that is associated with robots so as to make automation more salable to a broad audience can have unfortunate implications in terms of safety, especially as these larger scale environmental changes are occurring. Initiatives that foster the "social modulation" of fear of the "other" and related emotions in the realm of robot-related workplace contexts could play increasing roles in the expansion of robotics in organizations (Kavakli, 2019; Morozov & Ito, 2019), but can potentially result in mishaps. This chapter presents the case that efforts to infuse robotic entities in the workplace should be construed so that human participants retain some of their essential fear of autonomous entities, in this case, robots that are possibly dangerous in particular situations. Anxieties linked with automation extend beyond the prospects for death and injury, as automation has the prospects of displacing many kinds of occupations (McAfee & Brynjolfsson, 2016; Richardson, 2015).

The numbers of workplace deaths that are specifically attributable to robots and automated vehicles are often hard to obtain; information that specially targets robotics and AI as causal factors is sometimes not collected or subsequently provided to the public in the US. Data about such robot-inflicted casualties in other nations are generally treated with comparable indeterminacy. For many years, the US National Institute for Occupational Safety and Health (NIOSH) did not segregate such occurrences in its mortality databases, though it is beginning to do so:

> The National Institute for Occupational Safety and Health (NIOSH) has a Center for Occupational Robotics Research, and more specifically, a special program called the Fatality Assessment and Control Evaluation (FACE) Program. Through the FACE program, NIOSH is conducting robot safety surveillance, targeted investigations, and prevention activities. The program is conducting in-depth investigations of robot-related deaths. The FACE program is currently operated in 7 states through local state health or labor agencies. (ISIenvironmental, 2019, para. 2)

One of the worker deaths FACE is reportedly investigating is described in the following:

In one case, NIOSH investigated a death at a water bottling company where a worker was crushed. At the facility, each vehicle had safety sensors to detect objects or workers in the vehicle's path. An alarm would sound when an obstruction was present, and the vehicles would stop moving until the obstruction was removed.

The worker heard an alarm sound on one of the vehicles indicating sensors detected an object in its path. He attempted to remove a piece of plastic that likely tore off a pallet. Before removing the plastic and reaching under the forks, the worker had not cut the power to the machine. He also had not heeded label warnings to stay clear of the forks. Investigators believe that when he removed the plastic obstruction, he was positioned outside the path of the sensor. The vehicle resumed operation, the forks came down, and the worker was crushed. (ISIenvironmental, 2019, para. 4 and 5)

Although clarification of what was involved in a particular death is often conducted through courtroom activities that can last for decades, carefully-compiled rosters of deaths and injuries could indeed help in more timely ways in understanding what is going on in workplaces and communities. A compendium of negative incidents that involve AI-enhanced entities and robots is being crowdsourced by non-governmental agencies and individuals. Simonite (2021) characterizes this initiative in these terms:

> The AI Incident Database is hosted by Partnership on AI, a nonprofit founded by large tech companies to research the downsides of the technology. The roll of dishonor was started by Sean McGregor, who works as a machine learning engineer at voice processor startup Syntiant. He says it's needed because AI allows machines to intervene more directly in people's lives, but the culture of software engineering does not encourage safety. (para. 3)

The incidents accumulated through crowdsourcing or government agency efforts can aid in more than just assigning blame or determining legal liability. As related in Yampolskiy (2019), efforts to predicting future AI failures "can benefit from historic examples" (p. 138). Efforts to accumulate narratives about AI and robotic failures are framed by specific cultural contexts, with issues emerging worldwide in a broad assortment of settings (Gnambs & Appel, 2019; Kim et al., 2021; Nakada, 2019; Waytz & Norton, 2014; Xu & Yu, 2019).

Incidents concerning robots and automated vehicles are still attracting some news coverage, with the press in effect serving somewhat in a watchdog mode. For example, MacInnes (2021) described how Toyota temporarily paused its use of its Paralympics self-driving buses after one of the vehicles hit an athlete with visually impairments. In 2021, a Tesla that reportedly was under the control of Autopilot hit a police car that was parked near the side of a road (Isidore, 2021). A 2019 T crash in California that killed two individuals has triggered felony charges on the part of the driver because of alleged misuse of Autopilot (Moon, 2022). In India, a possible "snag in a sensor" reportedly resulted in the death by industrial robot of 44-year-old Umesh Ramesh Dhake in February 2021 (Express News Service, 2021). An assortment of recent accidents is associated with food delivery robots, often in which the robots are mistaken for animals by automobile drivers and bicyclists or are otherwise misperceived (Terbrack, 2021). At some point, the numbers and varieties of these robot and AI-related news reports may overwhelm or bore various international news consumers, and their dissemination will be less widespread, but for now these media accounts are serving as a de facto database for critical examination.

HISTORICAL AND SOCIAL PERSPECTIVES ON ROBOT- AND AI-INVOLVED DEATHS

As outlined in the first chapter, the International Federation of Robotics (IFR, 2019) defined a manufacturing robot as "an automatically controlled, reprogrammable and multipurpose manipulator for use in industrial automation applications," and provided comparable characterizations for service robots. The word "robot" is rooted in the 1920 Czech play *Rossum's Universal Robots* by Karel Čapek; the play (which featured human deaths by robot) gained international appeal when performed in the US in the 1920s (Abnet, 2020). The term stems from the Slavic term "robota," for forced-labor worker. Answers to the question of what is considered a robot have changed as automation has become more sophisticated; as related by Burdick (1992), "Once a certain level of automation becomes widespread, we no longer call it robotic. For example, 200 years ago a dishwasher would have been considered a robot" (p. 2). In the past decades, robots have taken on a wide variety of shapes and functions, and even include some human appendages such as mechanized arms for individuals whose limbs have been severed.

Human mortality and vulnerability have often been factors in how robotics and related artificial intelligence (AI) applications have been construed in public policy and legal venues, with Isaac Asimov's (1950) "Three Laws of Robotics" an early attempt to provide some direction (Clarke, 1993; Anderson, 2008). The autonomous entity HAL in *2001: A Space Odyssey* inspired widely-discussed questions such as Dennett's (1997) query, "Did HAL commit murder?" Linkages of robot-related problems to Mary Shelly's 1818 classic *Frankenstein* (Shelley, 2012) have continued to be strong, despite the fact that the monster involved was not mechanical (Abnet, 2020). In the past few years, a number of criticisms, legislative efforts, and vocal protests specifically directed at autonomous system implementations have also arisen, focusing on such technologies as drones, robots, self-driving cars, and online transportation services such as Uber (Ramirez, 2020). These criticisms are part of the backdrop and underpinnings for organizational implementation initiatives for autonomous entities, although they are often stifled in efforts to increase the acceptance of robots by workers. For example, the notion of "robophobia" has emerged to characterize real concerns of individuals and households about economic and physical damages (Akst, 2013). Robophobia is generally treated with psychological interventions targeted toward the individual rather than critical analysis and strategic initiatives to redesign organizational and community systems (Mays et al., 2021).

Deaths in workplace and community contexts are generally traumatic events that focus the attention of organizational participants as well as the community at large on certain public policy themes, such as gun control. Questions about "what happened" in particular incidents are often asked that are rooted in the interaction order of the settings involved (Goffman, 1983; Misztal, 2001), such as who or what is considered responsible for this deviation in normality or breach of trust. There are engineering, legal, moral, and social strategies for answering such questions in the realm of the automated workplace, although some of these are deficient in addressing needed concerns. For instance, from a technical perspective, Kirschgens et al. (2018) declare that "robot security is being ignored by manufacturers" (p. 1), and describes an assortment of emerging issues such as robot hacking. Ionescu, Schlund, and Schmidbauer (2019) outline comparable concerns rooted in current deficiencies in engineering knowledge and expertise in security areas. A growing academic literature is tackling specifically the topic of robot-related accidents and deaths (Bigman et al., 2019), with some theorists holding robots themselves increasingly accountable for such mishaps, linked to their expanding levels of autonomy (Asaro, 2007). As organizations find that the introduction of robots into the workplace is economically desirable, basic understandings about the integration of autonomous entities are being wrought. This chapter explores the notion that in order to help prevent deaths by robots and make the workplace more safe, the basic survival instincts of individuals should be utilized as a resource and not be overridden, diminished, or corrupted; robots are potentially akin to seemingly-benign wild animals who can kill their trainers, and the notion of "friendly and trusted companions" may increase short-run human–robot collaboration but be dysfunctional in the long run.

Issues of workplace deaths have critical implications for managers as well as workers; deaths that involve a component of autonomous, machine-originated action have broad implications for the insurance and other risk-related industries (Crootof, 2014; Lemley & Casey, 2019). A pioneering analysis of such deaths is by Goossens (1991): "Accidents with automatic production systems are reported to be on the order of one in a hundred or thousand robot-years, while fatal accidents are found to occur one or two orders of magnitude less frequently" (p. 221). The number and variety of workplace deaths in the US and some other developed nations are expanding, despite considerable resources directed toward implementation and safety efforts (US Bureau of Labor Statistics, 2019). Such accidents may be seen as inevitable, but safety measures have been shown to reduce their likelihood, though the expenses involved can increase employer reluctance to make the needed changes (Rogers, 2020). Employees sometimes have some options as to the kind of risks they face on the job: individuals who have high risk intolerance have been shown to gravitate toward employment with fewer opportunities for accidental injury or death (DeLeire & Levy, 2004).

Kinds of deaths that involve autonomous entities include death by industrial accident, as in the cases of Robert Williams and others described at the start of this chapter as well as in Winfield et al. (2020). Wildlife deaths are also generating attention: Gregory-Kumar (2021) and Rasmussen et al. (2021) describe how hedgehogs in household yards can be killed by the actions of robotic lawn robots, triggering the concern of animal rights activists. Other kinds of robot-involved death include death by autonomous vehicle (King, 2019), and death by medical instrument failure or accident (Tucker, 2018). Death with the involvement of robotic appendages and prostheses, or "killer robot arms," is another emerging venue (Gurney, 2018). Robot-related lethal incidents can also be on a large, societal scale as well as in the context of the individual. Crootof (2014) declares that "the killer robots are here" in everyday life as well as in larger international contexts. The lethal impacts of autonomous military robots are often discussed in the context of robot-involved deaths in the workplace (Skerker et al., 2020). The prospects of a "robotic apocalypse" (DiCarlo, 2016, p. 56) or an "AI Armageddon" (McCauley, 2007) have also generated and focused anxiety in this arena. For example, Sharkey (2012) has projected "the evitability of autonomous robot warfare" (p. 787), with robots the major figures in battles that ultimately kill human beings. In comparably alarmist terms, Sharkey et al. (2010) predict "the coming robot crime wave" (p. 115). Not all researchers and analysts have come to such dire conclusions. Christley (2015) posits that "Robots can't kill you – Claiming they can is dangerous," and George (2016) asserts that "No, the robots are not about to rise up and destroy us all." In an upcoming chapter, the notions of the "killer robot" and LAWS ("lethal autonomous weapons systems") that are developed and implemented by governmental forces will be unpacked, a level of organization a bit higher than the everyday, ubiquitous lethal robots and autonomous vehicles described in this chapter.

Robots as Artificial "Others" and "Thinking Machines"

The robot as an "other" in workplaces and communities has been a theme of science fiction and creative works of many varieties (Coeckelbergh, 2011; Higbie, 2013; Wennerscheid, 2019), with the robot often taking on negative roles including aggression and domination. As described by Bartneck et al. (2020), "Another archetypal story line is the robotic uprising. In short, humanity builds intelligent and strong robots, then the robots decide to take over the world and enslave or kill all humans" (p. 188). Analyses of the relative places of humans and machines in the workplace are not new, with public policy topics ranging from economic displacement to community safety (Pankewitz, 2017; Westerlund, 2020). Danzman and Colgan (2017) declare that "Robots aren't killing the American Dream," although Ramirez (2020) and Waytz and Norton (2014) present some evidence through economic data analyses that counters that claim.

The anthropomorphic dimensions of robots (such as a humanlike face) are sometimes an explicit aspect of design, and in other contexts are added features provided by users for entertainment (Corkery, 2020). Coeckelbergh (2011) describes how the linguistic construction of artificial others (such as addressing them by name) affects the way individuals construe them as entities. The perceived levels and kinds of robot intelligence are also factors in the kinds of human–robot interactions that emerge (Banks, 2019). The notion of robotic "superintelligence" has been seriously discussed in business and political discourse (Hurlburt, 2017). Other research on how robots can express dominance in interactions stresses robotic timing, placement, and even gender-related imagery (Peters, Broekens, Li, & Neerincx, 2019; Stein et al., 2020).

The notion of a "thinking machine" that can engage in functional and useful practices for workplaces as well as dysfunctional ones has permeated many robotics initiatives and AI approaches as well as public consciousness concerning their applications for decades, framing some of the ethical considerations involving these issues (Daugherty & Wilson, 2018). Such machines are often generally construed as autonomous intelligent entities independently operating on the world with various degrees of discretion and self-awareness; many projections have emerged through the decades of how these entities (including AI chatbots as well as more sophisticated and extensive systems) will increasingly permeate everyday life (Dennett, 1997; Tani, 2016). Thinking machines have played roles in science fiction novels and other entertainment genres, often in the form of intelligent and mobile robots. Some scenarios of the future project that humans themselves will be displaced in their basic functions by intelligent systems (Darling, 2016; Dihal, 2020). Other scenarios project that human beings will someday be able to upload aspects of themselves to these entities, in a form often construed as "artificial immortality" and the "posthuman divine" (DiCarlo, 2016; Ferrando, 2019), as described in a previous chapter. Possibly in preparation for this new era, robots who are considered "deceased" have been given funerals and eulogies by their human workplace colleagues (Carter et al., 2020). At Oregon State University, a funeral was recently given for a food service robot who was demolished while crossing a train track (Miller, 2022). Despite the assumed "absence of consciousness" of robots (Devillers, 2020, p. 261), the growing assortment of social and emotional robots are often associated in their existences with a kind of lifeworld that is close to that of humans (Dorner & Hille, 1995), and can possibly provide the robots with the capacity to understand the act of murder.

The thinking machine idea complicates and problematizes the issue of workplace deaths by robot. A machine that is thinking autonomously can be assigned a different, more responsible role in analyses of the deaths than one that is merely programmed to work in a particular, predictable manner. Battles between humans and machines for contested social spaces have often been projected, with the gaming terrains often constructed in ways that give autonomous entities an inherent advantage. For example, AI research has often been framed as experiencing successes in particular confrontations in gaming: chess-playing computers can defeat grand masters (with dozens of human beings working to assist the AI system's efforts), and systems such as IBM's *Watson* have taken on celebrity status in part because of particular successes in US game shows such as *Jeopardy* (Daugherty & Wilson, 2018). The thinking machine notion may indeed seem simple enough to be understood by a general audience but contains many assumptions and requires research and development underpinnings that involve advanced technological support (Westerlund, 2020). By framing automated systems in simplistic and generally positive terms that emphasize certain "wins" and downplay failures and setbacks, individuals can be disempowered who need to make sound decisions about robotic implementation and related safety matters.

Design and Implementation Issues Involving Robotics

Complete certainty is not possible in design and implementation efforts, and a certain number of tragic accidents can be expected in organizations whatever the levels of preparation. However, strategies for implementing "intrinsically safe" robots are being proposed (Pearlman, 2020), with the safety supposedly linked to the large extent of human–robot interaction and the related robot learning opportunities. Processes for the investigation of robot-induced fatalities can be handled in comparable ways to how airline-related deaths are handled (Winfield et al., 2020); currently, however, they are not, and much information is wasted by not providing intense analysis of every robot-inflicted death by teams of dedicated analysts. Crider (2017) proposes that data from accident investigations in robot-related fatalities be accumulated and shared nationally in much the same way the US National Transportation Safety Board

(NTSB) processes and analyzes information about airline mishaps. Hightech efforts in information technology have developed some approaches in accident investigation, but these may have to be extended dramatically when complex human–robot interactions, and the explainability issues of machine learning, are involved (Stowers et al., 2016). Many autonomous entities operate in ways that are not transparent in terms of their reasoning, with "machine learning" strategies rather than more straightforward and human decipherable programming. Some analysts worry that human operators will be assigned blame for human–robot accidents in lieu of undergoing thorough, systemwide examinations of what happened in the accident (Elish, 2019). Lack of sound accident protocols presents dangerous scenarios for the future: the prospects for "rogue robots" with problematic processes and poor security are growing, which may be at the root of some industrial mishaps (Maggi et al., 2017).

From a psychoanalytic perspective, the association of robotic entities with "magical" thinking (and related superstitious interactions) is salient, which can serve to complicate the human–robot relationship and related implementation issues:

> Magic control of technologies may engender the illusion of enhanced control on one's own emotional life. In particular, a narcissistic relationship with a robot may shield one from recognizing the other and one's own dependence from the other. (Scalzone & Tamburrini, 2013, p. 297)

Despite the uncertainty involved with integrating autonomous entities into the workplace (along with the "magical" thinking of some employees), many developers are working toward making the interactions between humans and robots more natural and intuitive, as if two humans were interacting rather than a human and a robot. In this perspective, increasing the levels of trust of humans for robots could make interactions more seamless and fluid (Spence et al., 2014). This can supposedly be accomplished by constructing the faces of robots so as to foster such trust (Kalegina et al., 2018). In order to deal with problematic or emergency situations, Briggs and Scheutz (2014) explored the effects of robotic displays of protest and distress on the humans with whom the robots are situated. Robinette, Wagner, and Howard (2016) investigated human–robot trust in emergency scenarios, while Strohkorb et al. (2018) examined the effects of a robot's perceived "vulnerable" behavior on trust in human–robot teams. The potential security flaws in many industrial robot configurations should put any human–robot trust in question, however (Kinzler et al., 2019; Wolfert, 2020). The technical and epistemic "debts" involved with many efforts to design automated environments (i.e., the deficits in security-related know-how to fend off potential attacks) are currently substantial, raising the potential for security breaches (Ionescu, Schlund, & Schmidbauer, 2019; Oravec, 2017).

Good design and adequate training are indeed expensive, and the kinds of efforts that can potentially improve human–robot environmental safety are often unfortunately neglected in many organizations and community settings (Rogers, 2020). Much of the communication about technological mishaps, cybercrime, and accident situations is informal, with individuals sharing science fiction themes and sketchy details rather than empirically-grounded findings (Wall, 2008). Legal remedies for robot-related accidents and deaths could be better defined in the near future (Lemley & Casey, 2019), with specific efforts toward increasing transparency of automated systems.

Human–Robot Collaboration Models for Safety

The robot- and AI-related human deaths described in this chapter often provide striking symbols of design failures, but also deficiencies in many organizations' training efforts. As with aviation disasters, every robot-related mishap is different, with variations based in the variations in security issues, environmental issues, and employee training. Much of the effort in building human–robot interaction is in making employees feel better about robots so they will be less fearful of the situation (Abnet, 2020). Training of humans so that they fit in more adequately with the perceived requirements of workplace operations has dimensions that include trust-building initiatives. Robots and automation as a whole are promoted in workplaces as a symbol of security and advancement (Beane, 2020), despite various technical deficiencies. However, the workplace presents a specific, confined setting in which employees must operate; if those employees (for whatever reason) are not afforded the means for ensuring their safety, no amount of psychological training or other influence will change the situation.

In industrial environments, the settings and circumstances in which humans and robots interact are becoming more intertwined, providing new opportunities for danger (Bragança et al., 2019). Humans in the workplace are often surveilled so that their behavior is apparently in consonance with workplace requirements for efficient and effective processes, and other interventions are being installed so as to support cultures of human-to-human trust and safety (Iversen & Rogers, 2020). Many workplaces are also being redesigned so that robots can function more efficiently, with the human being often construed as the alien in the robotic environment rather than the reverse (Miller, 1983). Such vehicles as wearable technologies, eye scanning, and webcams are being used to collect the employee data used for the surveillance initiatives that monitor the activities of the humans involved in human–robot interactions (Oravec, 2020). Social modulation of fear efforts can serve to make working with a robot appear less threatening (Morozov & Ito, 2019), regardless of the safety issues involved. For instance, some robots are being designed to read and respond to worker's moods and projected intensions so as to make collaboration smoother (Görür & Erkmen, 2019; Rani, 2004). Factors outside the workplace are also becoming salient in modulating the fear of robots; robotic pets and personal companions have made robots a familiar and less intimidating entity in many homes, schools, and communities (Melson et al., 2009).

There are some readily-available models to provide insights about human–robot interaction and collaboration. For example, humans have been co-existing with wild animals for millennia (Glikman, Frank, & Marchini, 2019), although human fatalities in these interactions are still a considerable factor. Human–robot relations have often been modeled in terms of human-animal relationships (Coeckelbergh, 2011; King, 2019). In 1862, US philosopher Henry David Thoreau published an essay that states "In wildness is the preservation of the world" (1991), but some of the wildness that is associated with both humans and machinery can be deadly. Lion trainers must realize that despite their best efforts to understand and befriend their lions, dangerous reactions can occur to certain stimuli. The kinds of human wildness that may be involved in preserving the world may indeed vary from robotic wildness, but both can play a role in maintaining healthy distances and separation between workers and autonomous workplace entities.

Moderating the characterizations of the capabilities of autonomous systems so that hyperbole is replaced by more specific and nuanced analyses may also be of help in developing solutions to the issues involved with robot-related accidents and deaths. This is not to say that technological vision is unimportant, nor that the motivation of researchers and developers should be neglected. Various game shows, contests, demonstrations, and widely-publicized chess matches have formed a good deal of the rhetorical framing of AI technologies; recent technological successes in some of these competitions have buttressed the claims of developers that certain kinds of automated systems should indeed be widely implemented, despite social costs, enactment complications, and moral concerns (Oravec, 2019). The projections and speculations about AI and robotics can be driven in measured ways by broader societal concerns and not primarily by individual or corporate-level opportunism.

The recommendations in the above paragraphs are not attempts to attack or diminish the technologies involved; uses of control systems and robotics have indeed empowered many individuals with disabilities and accomplished other socially-supportive objectives. However, a number of technological apologists have provided characterizations of AI-enabled autonomous systems that do little to elucidate critical aspects of the technologies and enable individuals to make well-considered decisions about their uses. As such applications as service robots and self-driving cars become omnipresent in everyday life, realistic and nuanced characterizations of these technologies are required in order for individuals and organizations to make sound decisions about safety.

Some Future Directions for Human–Robot-AI Interaction

Deaths and injuries associated with robots and other AI-enhanced entities are likely to continue; forthcoming research and development directions in this complex arena include examining the potential for anti-robot protests, terrorism, and abuse that are linked to the kinds of incidents previously described. Despite the many personal losses associated with automation through the decades (including deaths), worker protests associated directly against automation have been relatively few, with such recent exceptions as Uber-related demonstrations by taxi drivers and various forms of performance art (Swartz, 2015; Wright & Schultz, 2018). The prospects for such activity could be dramatically increased in the advent of the kinds of robot-inflicted deaths and injuries in workplaces described in this chapter; even a few deaths by autonomous cars reportedly precipitated a significant lowering of trust in such vehicles (Kaur & Rampersad, 2018). Survey research compiled by Gnambs and Appel (2019) shows growing unpopularity of robots in some contexts, generally rooted in the knowledge and awareness of real-world negative incidents. As related in previous sections, death and human mortality notions have had strong associations with robots for the past decades, from science fiction fantasies to the "uncanny valley" psychological associations with terror. Uncanny valley-related concepts stem from empirical research that shows that many individuals begin to recoil from robots as they become more similar to humans (Mori, 2012), as discussed in the first chapter. Personal privacy concerns are also arising as robots learn more exacting details about their human collaborators in order to accomplish various objectives (Lutz & Tamó-Larrieux, 2020), which could make particular death and injury incidents with robots both broadly relevant and more individualized in their significance. A death that might have occurred in part because personal details were taken into account (or were omitted for various reasons) can have chilling implications for other individuals who share those characteristics.

In past centuries, the Luddites were associated with anti-automation themes, with the term "luddite" often used as a generic reference to those who would attack technological implementations (Manuel, 1938). How far would employees and community participants go in protesting robot-inflicted injuries and deaths? Ringler and Reckter (2012) and Bartneck et al. (2007) explore the intentional destruction of robots, which has been demonstrated both in real-life workplaces and in experimental laboratory settings. A number of other research initiatives have followed on this theme, with Bartneck and Hu (2008) and Connolly (2020) examining conditions in which employees would abuse robots. With the allocation of blame to robots and automated entities has come discourse on potential punishments, with analyses by Lima et al. (2020) and Zając (2020). Whether destruction of a robot that has engaged in lethal activities would provide comfort to the families of the victims has yet to be determined, however.

Organizational transparency about robotic failures, artificial intelligence deficits, and related workplace concerns is a necessity to ensure that employees can establish a reasonable, grounded, and "guarded trust" in the robots with which they work as well as with other human organizational participants (Oravec, 2004). Such transparency may also decrease the potential for the anti-robot abuse and destruction discussed in detail in the following chapter; failures and problematic concerns can be better placed into context. Humans often generalize from experience dealing with other humans' behavior about the safety and utility of such interactions (Baer et al., 2018); individuals learn to trust each other by building relationships that span personal, social, and workplace factors. However, generalizing about robots in this way could indeed be fatal; every design and implementation situation is different when it comes to autonomous systems, with various manufacturers and developers having different standards and priorities. Fostering an atmosphere in which employees become aware of potential robot hazards, and be vigilant for new, emerging security concerns, can be a formidable organizational initiative (Wolfert et al., 2020); however, it is an effort that is essential for human safety.

Some Conclusions and Reflections

The unsettling prospects for deaths and injuries of humans at the hands of robots and other autonomous entities are increasingly influencing the character of modern workplaces, transportation systems, and communities. The memories of Wanda Holbrook, Robert Williams, Umesh Ramesh Dhake, Elaine Herzberg, and so many others killed by robots and autonomous vehicles will indeed continue to drive the efforts of safety specialists and workplace designers. These deaths in everyday contexts have real-life implications for individuals and their families in terms of pain and suffering, as well as economic and legal consequences for organizations as a whole; this chapter by no means is intended to make light of the damages and horrors involved. In workplace contexts, destructions or sabotages of robots by organizational participants (whether or not they are directly related to such deaths) could also produce substantial economic losses, and perhaps even precipitate collateral injury on the part of workers who inadvertently deal with hacked or partially-disabled robots. Although the numbers of individuals who smash robots and destroy autonomous entities are currently small, as the use of robots in workplaces continues and various anxieties compound the situation the possibility of acts of sabotage and even more substantial damages is increasing, as discussed in the next chapter.

Legal solutions are often limited for individuals who are injured by robots because of the diffuse responsibility for these occurrences (Asaro, 2007; Lemley & Casey, 2019). Multiple levels of designers, developers, and testers as well as human operators are potentially at legal fault for the operations of autonomous entities, and robot and autonomous vehicle architectures are often composed of parts or modules from different suppliers. The mottos and ethical statements of many high-tech organizations (including Alphabet/Google) often assert an intent to avoid "evil" and other societal troubles (Oravec, 2014); these robot-related developments could indeed present high-tech companies and workplace implementers with new moral and social challenges.

Current efforts to construct a robotic "other" in a workplace or community context have roots in the substantial legacy of robotics and AI over the past decades. Such a robotic other could indeed tighten bonds among individuals by emphasizing their humanity in contrast to their robotic colleagues; in short, people could be drawn together in their common humanity against robots. However, applications of robot-related research on how to redesign workplaces and educate workers for human–robot interaction can have a potentially manipulative quality as the robotic "other" is reframed and characterized in various ways. The social modulation of fear operates at many levels in workplaces; when fear is lessened to an appropriate extent, individuals can be enabled to engage in difficult and dangerous tasks alongside robots, but potentially with increased risk and lessened threat awareness. Whether or to what extent employee fears are buffered or facilitated should depend on health and safety considerations and not just economic concerns.

Despite the growing uncertainties involved with robot operations and security, robots are increasingly portrayed as non-problematic and even sympathetic companions and collaborators in many marketing and managerial efforts. Framing robots so as to be friendly colleagues in the confined settings of the workplace or community may help to increase their acceptance, but can potentially lead to deaths and injuries as individuals lose touch with their basic instincts concerning safety. Survival-related instincts of humans should not be dampened or overridden; the psychological distance between humans and robots can have survival value, both in the very specific case of the individual employee confronting a seemingly errant robot to the human society facing the potential of proliferation of lethal autonomous military robots.

REFERENCES

Abnet, D. A. (2020). *The American Robot*. University of Chicago Press.
Akst, D. (2013). Automation anxiety. *The Wilson Quarterly, 37*(3), 65–78.
Anderson, S. L. (2008). Asimov's 'three laws of robotics' and machine metaethics. *AI & Society, 22*(4), 477–493.

Asaro, P. M. (2007). Robots and responsibility from a legal perspective. *Proceedings of the IEEE, 4*(14), 20–24.
Asimov, I. (1950). *I, Robot*. Bantham Books.
Associated Press (2016, January 30). Company fined after worker's death in robotic machine. *Detroit News*. https://www.detroitnews.com/story/news/local/michigan/2016/01/30/company-fined-workers-death-robotic-machine/79562318/
Baer, M. D., Matta, F. K., Kim, J. K., Welsh, D. T., & Garud, N. (2018). It's not you, it's them: Social influences on trust propensity and trust dynamics. *Personnel Psychology, 71*(3), 423–455. https://doi.org/10.1111/peps.12265
Banks, J. (2019). Theory of mind in social robots: Replication of five established human tests. *International Journal of Social Robotics, 12*, 403–414. https://doi.org/10.1007/s12369-019-00588-x
Bartneck, C., Belpaeme, T., Eyssel, F., Kanda, T., Keijsers, M., & Šabanović, S. (2020). *Human-robot interaction: An introduction*. Cambridge University Press.
Bartneck, C., & Hu, J. (2008). Exploring the abuse of robots. *Interaction Studies, 9*(3), 415–433.
Bartneck, C., & Keijsers, M. (2020). The morality of abusing a robot. *Paladyn. Journal of Behavioral Robotics, 11*(1), 271–283.
Bartneck, C., Verbunt, M., Mubin, O., & Al Mahmud, A. (2007, March). To kill a mockingbird robot. In *Proceedings of the ACM/IEEE international conference on human-robot interaction*, pp. 81–87. https://doi.org/10.1145/1228716.1228728
Beane, M. I. (2020). In storage, yet on display: An empirical investigation of robots' value as social signals. In *Proceedings of the 2020 ACM/IEEE International Conference on Human-Robot Interaction*, pp. 83–91.
Berry, L. (2017). Mother of Alabama woman crushed to death by robot speaks as lawsuit looms. *Al.com*, https://www.al.com/business/2017/05/mother_of_alabama_woman_crushe.html
Bigman, Y. E., Waytz, A., Alterovitz, R., & Gray, K. (2019). Holding robots responsible: The elements of machine morality. *Trends in Cognitive Sciences, 23*(5), 365–368. https://doi.org/10.1016/j.tics.2019.02.008
Bragança S., Costa E., Castellucci I., & Arezes P. M. (2019). A brief overview of the use of collaborative robots in Industry 4.0: Human role and safety. In P. Arezes et al. (Eds.), *Occupational and Environmental Safety and Health. Studies in Systems, Decision and Control*, vol. 202. Springer. https://doi.org/10.1007/978-3-030-14730-3_68
Briggs, G., & Scheutz, M. (2014). How robots can affect human behavior: Investigating the effects of robotic displays of protest and distress. *International Journal of Social Robotics, 6*(3), 343–355. https://doi.org/10.1007/s12369-014-0235-1

Burdick, J. W. (1992). Robots that crawl, walk, and slither. *Engineering & Science, 55*(4), 2–13.

Carter, E. J., Reig, S., Tan, X. Z., Laput, G., Rosenthal, S., & Steinfeld, A. (2020, March). Death of a robot: Social media reactions and language usage when a robot stops operating. In *Proceedings of the 2020 ACM/IEEE International Conference on Human-Robot Interaction*, pp. 589–597. https://doi.org/10.1145/3319502.3374794

Cellan-Jones, R. (2020, September 16). Uber's self-driving operator charged over fatal crash. *BBC.com.* https://www.bbc.com/news/technology-54175359

Chrisley, R. (2015). Robots can't kill you: Claiming they can is dangerous. *The Conversation.* https://theconversation.com/robots-cant-kill-you-claiming-they-can-is-dangerous-44208

Clarke, R. (1993). Asimov's laws of robotics: Implications for information technology-Part I. *Computer, 26*(12), 53–61.

CNBC. (2021, September 3). Feds probe New York Tesla crash that killed man changing flat tire. https://www.cnbc.com/2021/09/03/feds-probe-new-york-tesla-crash-that-killed-man-changing-flat-tire-.html

Coeckelbergh, M. (2011). You, robot: On the linguistic construction of artificial others. *AI & Society, 26*(1), 61–69.

Connolly, J. (2020, March). Preventing robot abuse through emotional robot responses. In *Companion of the 2020 ACM/IEEE International Conference on Human-Robot Interaction*, pp. 558–560. https://doi.org/10.1145/3371382.3377433

Corkery, M. (2020, February 26). Should robots have a face? *New York Times.* https://www.nytimes.com/2020/02/26/business/robots-retail-jobs.html/

Courthouse News. (2017, March). Robot death. https://www.courthousenews.com/wp-content/uploads/2017/03/RobotDeath.pdf

Crider, D. A. (2017). The use of data from accident investigations in development of simulator training scenarios. In *AIAA Modeling and Simulation Technologies Conference*, p. 1078. https://doi.org/10.2514/6.2017-1078

Crootof, R. (2014). The killer robots are here: Legal and policy implications. *Cardozo Law Review, 36*, 1837–1915.

Danzman, S. B., & Colgan, J. D. (2017, March 10). Robots aren't killing the American dream. Neither is trade. This is the problem. *Washington Post.* https://www.washingtonpost.com/news/monkey-cage/wp/2017/03/10/robots-arent-killing-the-american-dream-neither-is-trade-this-is-the-real-problem/

Darling, K. (2016). *Extending legal protection to social robots: The effects of anthropomorphism, empathy, and violent behavior towards robotic objects.* Edward Elgar Publishing. https://doi.org/10.4337/9781783476732

Daugherty, P. R., & Wilson, H. J. (2018). *Human+ machine: Reimagining work in the age of AI.* Harvard Business Press.

DeLeire, T., & Levy, H. (2004). Worker sorting and the risk of death on the job. *Journal of Labor Economics*, *22*(4), 925–953.

Dennett, D. (1997). Did HAL commit murder? In David G. Stork (Ed.), *HAL's legacy: 2001's computer as dream and reality*, pp. 351–365. MIT Press. http://cogprints.org/430/

Devillers, L. (2020). Social and emotional robots: Useful artificial intelligence in the absence of consciousness. In *Healthcare and Artificial Intelligence*, pp. 261–267. Springer.

DiCarlo, C. (2016). How to avoid a robotic apocalypse: A consideration on the future developments of AI, emergent consciousness, and the Frankenstein Effect. *IEEE Technology and Society Magazine*, *35*(4), 56–61.

Dihal, K. (2020). Artificial intelligence, slavery, and revolt. *AI narratives: A history of imaginative thinking about intelligent machines*, 189–212.

Dorner, D., & Hille, K. (1995, October). Artificial souls: Motivated emotional robots. In *1995 IEEE International Conference on Systems, Man and Cybernetics. Intelligent Systems for the 21st Century*, *4*, 3828–3832. IEEE Press.

Elish, M. C. (2019). Moral crumple zones: Cautionary tales in human-robot interaction. *Engaging Science, Technology, and Society*, *5*, 40–60.

Express News Service. (2021, February 25). Indian Express. Injured critically in accident involving industrial robot, factory employee dies. https://indianexpress.com/article/cities/pune/injured-critically-in-accident-involving-industrial-robot-factory-employee-dies-7204970/

Ferrando, F. (2019). The posthuman divine: When robots can be enlightened. *Sophia*, *58*(4), 645–651.

Garber, M. (2013, September 20). Funerals for fallen robots. *The Atlantic*. https://www.theatlantic.com/technology/archive/2013/09/funerals-for-fallen-robots/279861/

George, D. (2016, January 23). No, the robots are not about to rise up and destroy us all. *World Economic Forum*. https://www.weforum.org/agenda/2016/01/no-the-robots-are-not-about-to-rise-up-and-destroy-us-all/

Glikman, J. A., Frank, B., & Marchini, S. (2019). Human-wildlife interactions: Multifaceted approaches for turning conflict into coexistence. *Human-wildlife interactions: Turning conflict into coexistence*, pp. 439–452.

Gnambs, T., & Appel, M. (2019). Are robots becoming unpopular? Changes in attitudes towards autonomous robotic systems in Europe. *Computers in Human Behavior*, *93*, 53–61.

Goffman, E. (1983). The interaction order: American Sociological Association, 1982 presidential address. *American Sociological Review*, *48*(1), 1–17.

Goossens, L. H. (1991). Risk prevention and policy-making in automatic systems. *Risk Analysis*, *11*(2), 217–228.

Görür, O. C., & Erkmen, A. M. (2019). Intention and body-mood engineering via proactive robot moves in HRI. In *Rapid automation: Concepts, methodologies, tools, and applications*, pp. 247–275. IGI Global.

Gould, H., Arnold, M., Kohn, T., Nansen, B., & Gibbs, M. (2021). Robot death care: A study of funerary practice. *International Journal of Cultural Studies, 24*(4), 603–621.

Gregory-Kumar, D. (2021, May 12). Hedgehog study to assess danger of robot lawn-mowers. *BBCNews.com*. https://www.bbc.com/news/uk-england-57084490

Gurney, D. (2018). Killer robot arms: A case-study in brain–computer interfaces and intentional acts. *Minds and Machines, 28*(4), 775–785.

Higbie, T. (2013). Why do robots rebel? The labor history of a cultural icon. *Labor: Studies in Working-Class History of the Americas, 10*(1), 99–121.

Hurlburt, G. (2017). Superintelligence: Myth or pressing reality? *It Professional, 19*(1), 6–11.

IFR, International Federation of Robotics. (2019). Industrial robot definition. https://ifr.org/#topics (accessed June 29, 2020).

Ionescu, T. B., Schlund, S., & Schmidbauer, C. (2019). Epistemic debt: A concept and measure of technical ignorance in smart manufacturing. In *International Conference on Applied Human Factors and Ergonomics*, pp. 81–93. Springer.

Isidore, C. (2021, August 30). Another Tesla reportedly using Autopilot hits a parked police car. *CNN.com*. https://www.cnn.com/2021/08/30/business/tesla-crash-police-car/

ISIenvironmental. (2019, August 29). Robot safety: NIOSH develops program to study robot-related injuries. https://isienvironmental.com/robot-safety-blog/

Iversen, C. M., & Rogers, A. (2020). Building a culture of safety and trust in team science. *Eos 101*, no. BNL-213859-2020-JAAM.

Kalegina, A., Schroeder, G., Allchin, A., Berlin, K., & Cakmak, M. (2018, February). Characterizing the design space of rendered robot faces. In *Proceedings of the 2018 ACM/IEEE International Conference on Human-Robot Interaction*, pp. 96–104. https://doi.org/10.1145/3171221.3171286

Kaur, K., & Rampersad, G. (2018). Trust in driverless cars: Investigating key factors influencing the adoption of driverless cars. *Journal of Engineering and Technology Management, 48*, 87–96. https://doi.org/10.1016/j.jengtecman.2018.04.006

Kavakli, M. (2019). Why do we have emotions? The social functions of emotions. *Research on Education and Psychology, 3*(1), 11–20.

Kim, S., Lee, J., & Kang, C. (2021). Analysis of industrial accidents causing through jamming or crushing accidental deaths in the manufacturing industry in South Korea: Focus on non-routine work on machinery. *Safety Science, 133*, 104998.

King, D. (2019). Putting the reins on autonomous vehicle liability: Why horse accidents are the best common law analogy. *North Carolina Journal of Law & Technology, 19*(4), 127–154.

Kinzler, M., Miller, J., Wu, Z., Williams, A., & Perouli, D. (2019, April). Cybersecurity vulnerabilities in two artificially intelligent humanoids on the market. In *Workshop on Technology and Consumer Protection (ConPro '19), held in conjunction with the 40th IEEE Symposium on Security and Privacy*. https://www.ieee-security.org/TC/SPW2019/ConPro/chapters/kinzler-conpro19.pdf

Kirschgens, L. A., Ugarte, I. Z., Uriarte, E. G., Rosas, A. M., & Vilches, V. M. (2018). Robot hazards: From safety to security. *arXiv preprint-arXiv:1806.06681*.

Lemley, M. A., & Casey, B. (2019). Remedies for robots. *The University of Chicago Law Review, 86*(5), 1311–1396.

Lima, G., Jeon, C., Cha, M., & Park, K. (2020, April). Will punishing robots become imperative in the future? In *Extended Abstracts of the 2020 CHI Conference on Human Factors in Computing Systems*, pp. 1–8.

Lutz, C., & Tamó-Larrieux, A. (2020). The robot privacy paradox: Understanding how privacy concerns shape intentions to use social robots. *Human-Machine Communication Journal* (HMC), *1*(1), 87–111. https://doi.org/10.30658/hmc.1.6

Lyons, S. (2018). *Death and the machine*. Palgrave Pivot.

MacInnes, P. (2021, August 27). Toyota pauses Paralympics self-driving buses after one hits visually impaired athlete. *The Guardian*, https://www.theguardian.com/technology/2021/aug/28/toyota-pauses-paralympics-self-driving-buses-after-one-hits-visually-impaired-athlete

Mays, K. K., Lei, Y., Giovanetti, R., & Katz, J. E. (2021). AI as a boss? A national US survey of predispositions governing comfort with expanded AI roles in society. *AI & SOCIETY, 1–14*,. https://doi.org/10.1007/s00146-021-01253-6

McAfee, A., & Brynjolfsson, E. (2016). Human work in the robotic future: Policy for the age of automation. *Foreign Affairs, 95*(4), 139–150.

McCauley, L. (2007). AI Armageddon and the three laws of robotics. *Ethics and Information Technology, 9*(2), 153–164.

Maggi, F., Quarta, D., Pogliani, M., Polino, M., Zanchettin, A. M., & Zanero, S. (2017). Rogue robots: Testing the limits of an industrial robot's security. *Trend Micro, Politecnico di Milano, Tech. Rep.* http://documents.trendmicro.com/assets/wp/wp-industrial-robot-security.pdf

Manuel, F. E. (1938). The Luddite movement in France. *The Journal of Modern History, 10*(2), 180–211.

Melson, G. F., Kahn, P. K., Beck, A., & Friedman, B. (2009). Robotic pets in human lives: Implications for the human–animal bond and for human relationships with personified technologies. *Journal of Social Issues, 65*(3), 545–567.

Miller, K. (2022, Winter). About those food robots. *Oregon Stater, 22*.

Miller, R. J. (1983). The human: Alien in the robotic environment? *The Annals of the American Academy of Political and Social Science, 470*(1), 11–15.

Misztal, B. A. (2001). Normality and trust in Goffman's theory of interaction order. *Sociological Theory, 19*(3), 312–324.

Moon, M. (2022, January 19). Tesla driver in fatal California crash first to face felony charges involving Autopilot. *Engadget*. https://www.engadget.com/tesla-driver-california-crash-first-to-face-felony-charges-autopilot-073720822.html

Mori, M. (2012, June 12). The uncanny valley: The original essay by Masahiro Mori. *IEEE Robots & Automation Magazine*, pp. 98–100. https://ieeexplore.ieee.org/xpl/tocresult.jsp?isnumber=6213218&punumber=100

Morozov, A., & Ito, W. (2019). Social modulation of fear: Facilitation vs buffering. *Genes, Brain and Behavior, 18*(1), e12491.

Nakada, M. (2019). Robots seen from the perspectives of Japanese culture, philosophy, ethics and *Aida* (betweenness). In T. Lennerfors & K. Murata (Eds.), *Tetsugaku Companion to Japanese Ethics and Technology. Tetsugaku Companions to Japanese Philosophy*, vol. 1. Springer. https://doi.org/10.1007/978-3-319-59027-1_8

Nunes, A., & Axhausen, K. W. (2021). Road safety, health inequity and the imminence of autonomous vehicles. *Nature Machine Intelligence, 3*(8), 654–655. https://doi.org/10.1038/s42256-021-00382-3

Oravec, J. A. (2004). Trusting in others' biases: Fostering guarded trust in collaborative filtering and recommender systems. *Knowledge, Technology & Policy, 17*(3–4), 106–123.

Oravec, J. A. (2014). Mottos and ethical statements of Internet-based organizations: Implications for corporate social responsibility. *International Journal of Civic Engagement and Social Change (IJCESC), 1*(2), 37–53.

Oravec, J. A. (2017). Kill switches, remote deletion, and intelligent agents: Framing everyday household cybersecurity in the internet of things. *Technology in Society, 51*, 189–198.

Oravec, J. A. (2019). Artificial intelligence, automation, and social welfare: Some ethical and historical perspectives on technological overstatement and hyperbole. *Ethics and Social Welfare, 13*(1), 18–32.

Oravec, J. A. (2020). Digital iatrogenesis and workplace marginalization: Some ethical issues involving self-tracking medical technologies. *Information, Communication & Society, 23*(14), 2030–2046.

Pankewitz, C. (2017). Automation, robots, and algorithms will drive the next stage of digital disruption. In *Phantom Ex Machina*, pp. 185–196. Springer.

Pearlman, L. (2020). Workplace safety in the roaring twenties. *EHS Today, 13*(5), 10–11.

Peters, R., Broekens, J., Li, K., & Neerincx, M. A. (2019, September). Robots expressing dominance: Effects of behaviours and modulation. In *2019 8th International Conference on Affective Computing and Intelligent Interaction (ACII)*, pp. 1–7. IEEE Press. https://doi.org/10.1109/ACII.2019.8925500

Ramirez, J. (2020). *Against automation mythologies: Business science fiction and the ruse of the robots*. Routledge.

Rani, P., Sarkar, N., Smith, C. A., & Kirby, L. D. (2004). Anxiety detecting robotic system–towards implicit human-robot collaboration. *Robotica, 22*(1), 85–95. https://doi.org/10.1017/S0263574703005319

Rasmussen, S. L., Schrøder, A. E., Mathiesen, R., Nielsen, J. L., Pertoldi, C., & Macdonald, D. W. (2021). Wildlife Conservation at a Garden Level: The Effect of Robotic Lawn Mowers on European Hedgehogs (Erinaceus europaeus). *Animals, 11*(5), 1–13. https://doi.org/10.3390/ani11051191

Richardson, K. (2015). *An anthropology of robots and AI: Annihilation anxiety and machines*. Routledge.

Ringler, J., & Reckter, H. (2012). DESU 100: About the temptation to destroy a robot. In *Proceedings of the Sixth International Conference on Tangible, Embedded and Embodied Interaction*, pp. 151–152.

Robinette, P., Wagner, A. R., & Howard, A. M. (2016). Investigating human-robot trust in emergency scenarios: Methodological lessons learned. In *Robust Intelligence and Trust in Autonomous Systems*, pp. 143–166. Springer. https://doi.org/10.1007/978-1-4899-7668-0_8

Rogers, B. (2020). The law and political economy of workplace technological change. *Harvard CR-CLL Review, 55*, 531–584. https://doi.org/10.2139/ssrn.3327608

Sahni, I. (2015, August 16). Manesar: Factory worker crushed to death by industrial robot. *Hindustan Times*. https://www.hindustantimes.com/gurgaon/manesar-factory-worker-crushed-to-death-by-industrial-robot/story-0Hc7V2uu2L2jlYfo9gEdXK.html

Salvini, P., Paez-Granados, D., & Billard, A. (2021). Safety concerns emerging from robots navigating in crowded pedestrian areas. *International Journal of Social Robotics*. https://doi.org/10.1007/s12369-021-00796-4

Scalzone, F., & Tamburrini, G. (2013). Human-robot interaction and psychoanalysis. *AI & Society, 28*(3), 297–307.

Sharkey, N. E. (2012). The evitability of autonomous robot warfare. *International Review of the Red Cross, 94*, 787–799.

Sharkey, N., Goodman, M., & Ross, N. (2010). The coming robot crime wave. *Computer, 43*(8), 115–116.

Sharma, M. D. (2021). *Top inspiring thoughts of Mark Zuckerberg.* Prabhat Prakashan.

Shelley, M. (2012). *Frankenstein.* Broadview Press.

Simonite, T. (2021, June 3). Don't end up on this artificial intelligence Hall of Shame. *Wired.* https://www.wired.com/story/artificial-intelligence-hall-shame/

Skerker, M., Purves, D., & Jenkins, R. (2020). Autonomous weapons systems and the moral equality of combatants. *Ethics and Information Technology, 22*(3), 197–209.

Spence, P. R., Westerman, D., Edwards, C., & Edwards, A. (2014). Welcoming our robot overlords: Initial expectations about interaction with a robot. *Communication Research Reports, 31*(3), 272–280.

Ssanderson, L. M., Collins, J. W., & McGlothlin, J. D. (1986). Robot-related fatality involving a US manufacturing plant employee: Case report and recommendations. *Journal of Occupational Accidents, 8*(1–2), 13–23.

Stein, J. P., Appel, M., Jost, A., & Ohler, P. (2020). Matter over mind? How the acceptance of digital entities depends on their appearance, mental prowess, and the interaction between both. *International Journal of Human-Computer Studies, 142,*. https://doi.org/10.1016/j.ijhcs.2020.102463

Stowers, K., Leyva, K., Hancock, G. M., & Hancock, P. A. (2016). Life or death by robot? *Ergonomics in Design, 24*(3), 17–22. https://doi.org/10.1177/1064804616635811

Strohkorb Sebo, S., Traeger, M., Jung, M., & Scassellati, B. (2018, February). The ripple effects of vulnerability: The effects of a robot's vulnerable behavior on trust in human-robot teams. In *Proceedings of the 2018 ACM/IEEE International Conference on Human-Robot Interaction*, pp. 178–186. https://doi.org/10.1145/3171221.3171275

Swartz, J. (2015, March 14). Protesters stage anti-robot rally at SXSW. *USA Today.* https://www.usatoday.com/story/tech/2015/03/14/sxsw-robot-ai-protest-artificial-intelligence/24777871/

Tani, J. (2016). *Exploring robotic minds: Actions, symbols, and consciousness as self-organizing dynamic phenomena.* Oxford University Press.

Terbrack, J. (2021, March 30). Starship vs. whips: Robots pose potential risk on streets. *Bowling Green Falcon Media.* https://www.bgfalconmedia.com/news/starship-vs-whips-robots-pose-potential-risk-on-streets/article_f31331b2-91af-11eb-af2b-d7feb69985a3.html

Thoreau, H. D. (1991). Walking. *The Great New Wilderness Debate*, pp. 31–41.

Tucker, I. (2018, March 25). Death by robot: The new mechanised danger in our changing world. *The Guardian.* https://www.theguardian.com/technology/2018/mar/25/death-by-robot-mechanised-danger-in-our-changing-world

UPI. (1981, December 8). Robot kills man. *United Press International.* https://www.upi.com/Archives/1981/12/08/Robot-kills-man/2127376635600/

US Bureau of Labor Statistics. (2019, December). *Economic news release.* https://www.bls.gov/news.release/cfoi.nr0.htm

Wakabayashi, D. (2018, March 19). Self-driving Uber car kills pedestrian in Arizona, Where robots roam. *New York Times.* https://www.nytimes.com/2018/03/19/technology/uber-driverless-fatality.html

Wall, D. S. (2008). Cybercrime and the culture of fear: Social science fiction(s) and the production of knowledge about cybercrime. *Information, Communication & Society, 11*(6), 861–884.

Watt, M. C., Maitland, R. A., & Gallagher, C. E. (2017). A case of the "heeby jeebies": An examination of intuitive judgements of "creepiness." *Canadian Journal of Behavioural Science/revue Canadienne Des Sciences Du Comportement, 49*(1), 58.

Waytz, A., & Norton, M. L. (2014). Botsourcing and outsourcing: Robot, British, Chinese, and German workers are for thinking—not feeling—jobs. *Emotion, 14*(2), 434–444.

Wennerscheid, S. (2019). Not in the image of humans: Robots as humans' other in contemporary science fiction film, literature and art. In *The transhumanism handbook* (pp. 557–571). Springer.

Westerlund, M. (2020). The ethical dimensions of public opinion on smart robots. *Technology Innovation Management Review, 10*(2), 25–36. https://doi.org/10.22215/timreview/1326

Winfield, A. F., Winkle, K., Webb, H., Lyngs, U., Jirotka, M., & Macrae, C. (2020). Robot Accident Investigation: A case study in Responsible Robotics. *arXiv preprint arXiv:2005.07474.*

Wolfert, P., Deschuyteneer, J., Oetringer, D., Robinson, N., & Belpaeme, T. (2020, March). Security risks of social robots used to persuade and manipulate: A proof of concept study. In *Companion of the 2020 ACM/IEEE International Conference on Human-Robot Interaction,* pp. 523–525.

Wright, S. A., & Schultz, A. (2018). The rising tide of artificial intelligence and business automation: Developing an ethical framework. *Business Horizons, 61*(6), 823–832.

Xu, L., & Yu, F. (2019). Factors that influence robot acceptance. *Chinese Science Bulletin, 65*(6), 496–510.

Yampolskiy, R. V. (2019). Predicting future AI failures from historic examples. *Foresight, 21*(1), 138–152. https://doi.org/10.1108/FS-04-2018-0034

Zając, M. (2020). Punishing robots–Way out of Sparrow's responsibility attribution problem. *Journal of Military Ethics, 19*(4), 285–291.

CHAPTER 8

Robo-Rage Against the Machine: Abuse, Sabotage, and Bullying of Robots and Autonomous Vehicles

Organizations and communities in modern societies have often dealt with forms of human-to-human aggression and abuse and are recently confronting situations involving human confrontations with AI-enhanced entities. Expressions of anti-robot aggression and related security breaches have apparently increased, despite extensive anthropomorphic research on how to make robots more palatable and attractive in particular contexts (Bankins & Formosa, 2020; Bartneck & Keijsers, 2020; Whitby, 2008). As humans are forced to work with and be monitored by robots, such incidents may expand in number and variety; as the compulsory utilization of autonomous vehicles becomes a factor, aggression against these entities could also multiply. The notion of humans "being treated like a robot" has itself been associated with images and narratives of abuse for decades, sometimes linked with human slavery (Hampton, 2015).

This chapter analyzes emerging varieties of robot and autonomous entity sabotage, bullying, manipulation, and destruction by humans, placing them in context of security concerns as well as other kinds of aggression and abusive conduct. Why would anyone attack a robot or self-driving vehicle? An increasing amount of this behavior can be considered in terms of "self-defense," given the specific intrusions of robots and drones in personal spaces (Froomkin & Colangelo, 2015). However, other instances are more specifically aggressive expressive movements, in dramaturgical terms being "frontstage" actions either with human or

© The Author(s), under exclusive license to Springer Nature Switzerland AG 2022
J. A. Oravec, *Good Robot, Bad Robot*, Social and Cultural Studies of Robots and AI, https://doi.org/10.1007/978-3-031-14013-6_8

robotic entities as audiences. Taking into account the intelligent functioning of robots (as discussed later in this chapter), some of the abuse could be characterized as in keeping with robot-as-personage and others more directed toward robot-as-property themes. The chapter also puts robot-related violence into perspective with various kinds of social rebellions and protests, exploring how attacks against autonomous entities (or disseminated narratives and images of the attacks) apparently often relate to externalizations of the economic and symbolic violence endured by individuals. A dramaturgical perspective can also be useful here: people who lose their livelihoods often act out on a public stage against what they may feel is the proximate cause of their losses, even if it is unclear as to what were the causes of their job loss or other personal distress.

There is plenty to act out against for humans. The previous chapter outlined how the deaths and injuries inflicted by robots and other AI-enhanced entities are unsettling many individuals. The presence of robots and other autonomous entities in workplaces, community settings, and households is creating new and often-foreboding types of environments and expanding the range of behaviors of humans in those spheres. Some of these human responses can indeed be healthy and useful, given particular circumstances, such as complaining to city councils about the use of robots to patrol city streets (While et al., 2021); others are clearly in the realm of "acting out" and involve dysfunction, human injury, and robotic destruction. With social media recordings and narratives, new spectators for these activities are often involved; as many of these activities are recorded for managerial or police intervention, other kinds of audiences also emerge. The confined quarters of the workplace can be the scene of violence, whether directed toward personages or property. Workplace violence of many varieties has had serious consequences for organizations and communities in the past decades (Neuman & Baron, 1998; Turpin et al., 2019), from spectacular mass shootings to surreptitious machine sabotage. The roles that attacks of robots and autonomous entities may play in the cumulative impacts of workplace violence are as yet emerging, but merit close attention given the growing societal importance and extensive permeation of automation. As robots and other AI-enhanced entities enter the home, the abuse of these technologies as a facet of household disruption and even domestic violence can also become an issue (PenzeyMoog & Slakoff, 2021).

With social media and other forms of online attention, even a few video or narrative portrayals of anti-robot attacks can have enlarged influences as they engender discourse about the place of AI and automation in society (Küster et al., 2021; Zeller et al., 2020). For example, parody videos of a Boston Dynamics robot being kicked (presumably in a display of its stability) became viral on several social media platforms, stimulating discourse on whether such expressions of violence should be allowed online (Moses & Ford, 2021). Mattoni and Teune (2014) present from an historical perspective how images disseminated through social media have played influential roles in social disruptions, sometimes in ways not directly intended or foreseen by those whose actions are portrayed in the images. Will violence against robots and autonomous vehicles become commonplace, and varieties of specially-targeted "hate crimes" against these entities emerge? Although attacks of robots and autonomous vehicles are currently not as frequent as many other kinds of aggression (such as domestic violence and worker-to-worker violence), the overall societal impacts of the attacks may be considerable, as well as the future instances of violence potentially linked to them.

As outlined in this book, automation-related changes in workplaces and communities have engendered an assortment of social anxieties, with different impacts in various regions worldwide, so cultural and national differences can complicate analyses of robot abuse (Gorgan, 2019; Xu & Yu, 2019). For example, from a UK setting, Payne provides a characterization of robot abuse as distracting but not overly aggressive: "We heard stories of workers standing in the way of robots, and minor acts of sabotage – and not playing along with them" (quoted in Bernal, 2019, para. 4). In contrast, from a South African context, Rodell (2020) asserts "We burn our robots in Africa; they keep trying to steal our jobs" (para. 1). In Moscow, an attack of a robot on the street was identified as a kind of "road rage" incident; the robot Alantim was "hit with a baseball bat by a thug, who eventually kicks the robot down to the ground…[it] is heard saying 'Pomogite', Russian for 'Help me'" (Chapman, 2017, para.1). Age differences can also be a factor in analyses, with numerous widely-publicized robot bullying cases involving children triggering concerns in Japan (Nomura et al., 2016) as well as in the US (Pearson & Borenstein, 2021). The potential for vicarious and voyeuristic appreciation of attacks upon robots and autonomous entities may expand as coverage of robot or autonomous vehicle abuse in popular and social media increases worldwide and possible "copycat" incidents or incident clusters occur (such as those described in the next section). Survey research compiled by Gnambs and Appel (2019) shows growing unpopularity of robots in many settings.

Efforts to analyze anti-robot activity are also complicated by time factors; technologies in the autonomous system arena advance quickly along with shifts in robot capabilities and subsequent human responses. The robot that is despised today may not be around tomorrow in a fast-paced automation environment. For example, the humanoid robot Pepper had a significant presence in social robots development and usage since 2014 (Aaltonen et al., 2017). However, production of the robots was halted in 2021 by SoftBank (Hyuga & Alpeyev, 2021). Individuals who might not have liked this robot because of experiences with it or some of its surface characteristics may have to find new targets for their anti-robot activities.

Mistreatment of Technologies in Everyday Workplace and Community Settings

Various forms of abuses of technological entities by humans are developing as pervasive features of many modern workplaces and community settings, with drones and autonomous vehicles joining robots in this regard (Lloyd & Payne, 2019; Mars, 2019). For this chapter, "abuse" constitutes non-intended and dysfunctional uses of an entity or artifact in a particular system context, with an emphasis on its destruction, debilitation, or desecration; a taxonomy of robot and autonomous vehicle abuses is provided in a section to follow. The intensity of the abuse can indeed vary dramatically, from nonchalant "impoliteness" (as described in robotic interactions by Rehm & Krogsager, 2013), to desecration, and to full-scale destruction.

Targeted and intentional destruction of workplace machinery has a long legacy. The "machine wrecking" and "machine smashing" in workplaces in previous decades provide some insights about automation-related abuses (Pearson, 1979; Tierney, 2019). In a past century (approximately 1815–1848), the Luddites' aggressive activities drew the attention of political and social leaders to particular issues. The Luddites were skilled machinery workers who chose to counter certain technological shifts (Linton, 1992). These followers of Ned Ludd were subsequently associated in historical contexts with anti-automation themes, with the term "luddite" often used in reference to such perspectives (Dorson, 1965; Jones, 2013; Manuel, 1938). Machine wrecking notions often stop short, however, in characterizing the damage occurring to the current set of autonomous entities. The destruction, manipulation, or sabotage of a thinking entity can have different dimensions than comparable efforts on a machine that is merely performing a routine and predictable set of functions. For example, autonomous machines can be manipulated and reprogrammed in their decision-making efforts (creating a "rogue" entity), complex maneuvers that are generally not applicable to non-intelligent machines. The type of abuse involved in any particular incident can differ depending on the perspectives of those viewing and interpreting the abuse, with external audiences' analyses and disseminations of the recorded or narrated event providing even more complications.

As analyzed in this chapter, activities that negatively affect robots and autonomous vehicles can have significant parallels with human-to-human abuse, bullying, or gaming in part because of the intelligent aspects of the artifacts involved. For instance, the bullying (repeated and targeted abuse) of robots or chatbots may have comparable features with human-to-human bullying as the autonomous entities are designed to interact in a humanlike fashion (Bartneck & Keijsers, 2020; D'Cruz & Noronha, 2021; Salvini et al., 2010). The larger social implications of the publicity of attacks against robots and autonomous entities are often comparable to those of human-to-human aggression as the incidents stimulate public discourse, focus organizational attention on human–robot interactions, and even stimulate copycat actions. Consider the following account of robot abuse:

> Every day for 10 months, Knightscope K5 patrolled the parking garage across the street from the city hall in Hayward, California. An autonomous security robot, it rolled around by itself, taking video and reading license plates. Locals had complained the garage was dangerous, but K5 seemed to be doing a good job restoring safety. Until the night of August 3, when a stranger came up to K5, knocked it down, and kicked it repeatedly, inflicting serious damage. (Harrison, 2019, para. 1)

Such abuse of the Knightscope security robots has reportedly continued in many communities, with "few tangible results" in countering crime (Farivar, 2021). The travels of hitchBOT provide another example of possible anti-robot fervor (Fraser et al., 2019). The autonomous, mobile robot hitchBOT successfully navigated lengthy ventures in Canada without a human escort, but was destroyed by individuals it encountered in the US after only a short time in transit. The widely-publicized stories of K5 and hitchBOT are coupled with many other accounts of the manipulation and sabotage of advanced technologies, with anthropomorphism often adding the dimensions of human vengeance or antipathy to the narratives (Harrison, 2019; Kolodny, 2018).

Many other instances of robot-related abuse and destruction are emerging in industrial, security, and service environments, often endangering nearby individuals as well as the robot in question (Black, 2019; Krumins, 2017; Luria et al., 2020b; Sherman, 2018; Torrez, 2019; Winfield et al., 2020), despite managerial and developer efforts to make the robots involved more appealing to humans. For example, mobile food service robots are often physically "kicked" by passers-by, sometimes with an attempt to "rip the flag" from the robots' antennas (Hamilton, 2018); some individuals are arguing against implementing them in certain contexts in part because of such destructive occurrences (Cox, 2020; Lynn, 2020). Security breaches involving the manipulation of programming are another form of attack (such as reprogramming a mobile food robot to engage in dangerous traffic maneuvers), which can be less immediately visible to outsiders but have dramatic consequences for the robots' operations as well as for the safety of human bystanders (Perales Gómez et al., 2021). Public attention to robot abuse issues also increased with reports that a sex robot was heavily damaged at an electronics show where it was on display (Nichols, 2017).

Along with robots, various forms of human attacks of autonomous vehicles (such as self-driving cars and trucks) have also been recently documented (Eliot, 2019; Liu et al., 2020; Moore et al., 2020); such incidents can produce collateral human damages as the vehicles hit bystanders or jeopardize their passengers' wellbeing. Serious questions are emerging of whether this abuse will present challenges in everyday driving contexts as human-controlled vehicles begin to share the road with self-driving cars. For example, Sabur (2018) reports that "according to the California Department of Motor Vehicles (DMV), the robotic car's mere presence prompted a man to run across the street against the do not walk symbol, shouting, and [strike] the left side of the Cruise AV's rear bumper and hatch with his entire body …." (para. 4). Numerous reports about Arizona residents attacking self-driving vehicles while "wielding rocks and knives" and related incidents have attracted both popular attention and the concern of manufacturers (Hamilton, 2019; Lee, 2020; Liu et al., 2020; Randazzo, 2018). In May 2021, the manipulation of autonomous vehicles to simulate that human co-drivers were monitoring them while on the road was a major topic of popular discourse in the US and UK, with this strategy reportedly resulting in accidents (Levin, 2021; Tangermann, 2021). Drones have comparably been destroyed in the air by individuals firing weapons or sent off course in dysfunctional ways (Cauchard et al., 2016; Matyszczyk, 2015).

Many incidents in which individuals express their anger or aggression with their control of vehicles (whether or not they have AI-enhanced features) have been documented (DiFruscia, 2012). "Road rage" occurrences of various sorts can indeed have negative outcomes on the individuals directly involved or on bystanders (Condliffe, 2016; Culver, 2018; Stephens et al., 2016); incidents on bicycles (or "cycle rage") can also have negative consequences as individuals' attacks are enhanced with the force of their vehicles (Lloyd, 2017). However, many of the emerging forms of abuse and sabotage of autonomous vehicles are apparently of a variety that impedes or disrupts the autonomous entities' decision-making functioning, based on some prior knowledge or assumptions about their operations (for instance, frustrating the entity in completion of its objectives). Strategies for engaging in such gaming and manipulation are already being actively crowdsourced online (Moore et al., 2020). Many recent attacks on autonomous entities also despoil or attack the dignity of the vehicle in some way, for example, by defacing it or engaging with it in a mocking or manipulative manner (Terbrack, 2021). These autonomous entities can be fragile in some respects (with delicate sensors, for instance) and apparently minor tampering can have substantial impacts on the entities' operational capacities, outcomes that might not immediately be seen but can emerge later in further operations. As new varieties of attacks of entities that have intelligent components emerge, new ways of responding or retaliating to aggression are also developing (as characterized in an upcoming section), including strategically-organized bystander activities by humans as well as other robots (Connolly et al., 2020). The potentials for "hate crime" or "hate incident" characterizations for some anti-robot activities are growing. In hate crimes and incidents, "the perpetrator is motivated by hostility or prejudice toward protected characteristics of the victim" (Bacon, May, & Charlesford, 2021), with the term "incident" used when the attack does not meet the standards for specific crimes. Hate crime approaches can foster discourse on the motivations of attackers (anti-robot abhorrence, for example) as well as on the societal contexts of the attacks.

Aggressive Anti-Robot Activity as Response to Symbolic Violence

Activities that involve physical damage (such as smashing a window) can often be seen as part of protest or social expression initiatives; damages of this sort are generally not excused in ethical terms by their association with political or social protests, but their meanings can be better deciphered if the attacks are placed in context of such expression (Joosse, 2020; Oravec, 2017a). Technological developments are often a theme in such protests, including genetically-modified foods and nuclear power (LeVasseur, 2017; Sale, 1996; Walsh, 1986). Other kinds of technologically-themed confrontations are emerging: for instance, Smith (2020) describes how more than fifty 5G cell towers were vandalized in the UK, with the destruction reportedly related to the proliferation of Coronavirus-related theories. Labor protests involving the implementation of robots have erupted at Rotterdam and Los Angeles ports (Morris, 2015; Roosevelt, 2019) as well as Las Vegas hotels (Oppenheimer, 2018). Nevett (2018) describes how the robot sex brothel Lumidolls had to be moved because of the protests of sex workers. Concerns about the relative places of humans and machines in the workplace and the economy at large have been disseminated for decades, with themes ranging from economic displacement to public safety (Hong, 2004; Pankewitz, 2017; Westerlund, 2020). According to Czarniawska and Joerges (2020), "'Robots could take half of the jobs in Germany' is a typical newspaper's title nowadays" (p. 2). Some organizations have reportedly attempted to expand anxieties concerning robots in order to gain opportunistic edges against workers in labor negotiations (Ramirez, 2020; Waytz & Norton, 2014). In such contentious contexts, violent and aggressive behaviors toward particular objects are often forms of disruption that can have wider implications as their theatrical and symbolic impacts are manifested (Graver, 1995; Tavory & Fine, 2020).

Kinds of economic perils as well as "symbolic violence" (Bourdieu, 1991) relating to recent employment losses have expanded in the advent of autonomous entities (Bernhardtz, 2019; Bharadwaj & Dvorkin, 2019), along with expressions of anxiety and disorientation by industrial and transportation workers (Kracher, 2005; MacKinnon, 2019; Sainato, 2020). Crowley (2014) describes the "relational indignity" that can result when new functional relationships and social standings are imposed on a particular setting. Many workplace incivilities are apparently rooted in conflicts of interest and group-level affronts, including physical damage as well as human-to-human conflicts (Roscigno et al., 2009). Keijsers and Bartneck (2018) contend that "dehumanisation occurs in human–robot interaction and like that in human–human interaction, it is linked to aggressive behavior" (p. 205). Current and potential robot-related symbolic violence involves the overall impacts of changes on individuals faced with technological shifts in workplaces and communities (Weininger, 2002). In turn, the notions of "destruction" and "disruption" in relation to the kinds of changes associated with automation are often met with positive responses in the managerial and economic literatures (e.g., Bergek et al., 2013), but with less welcoming responses on the part of some other organizational participants (Grint & Case, 1998).

Many of today's robots and autonomous vehicles have fragile and brittle dimensions that can be exploited or damaged in some way, although efforts are underway to "harden" them to be more resilient against attacks (Alexis, 2020). Destructive activities involving such entities are found within wider contexts of deviant behaviors in workplaces that are costing employers billions of dollars and often jeopardizing the lives of employees (Singh, 2020; Yeşiltaş & Gürlek, 2020). Images of intentionally-arranged robot destruction are readily found (Black, 2019), with many robot-to-robot battles placed on YouTube (and sometimes deleted by YouTube administrators, reportedly because of resemblance to animal battles). Lewnard (2020) describes such a setting in an educational context: "The Prospect High School Fieldhouse looked like a scaled-down scene out of a "Mad Max" movie, with marauding robots designed for one task – destruction" (para. 1). Numerous science fiction accounts involve robotic demise, often in activities with medieval themes of jousting and direct confrontation as well as more modern configurations (Luria et al., 2020a).

Consider this description of student "robot club" activities:

> These days, robots are used for everything from manufacturing cars, to exploring space… This April, a team of Rochester Institute of Technology undergrads will attempt to use a robot in a different way — specifically, to bash, batter, cut, puncture or burn a variety of other, similarly hostile machines into oblivion… *BattleBots* tasks participants with designing, building and piloting remotely controlled robotic gladiators that face-off in a hazard-filled arena. The rules are simple: The robots battle it out until one is disabled or destroyed. Each robot is equipped with its own weapon

system, which can range from spinning saws, to whirling bludgeons, to mechanical jaws, to flamethrowers. Losers are eliminated from the tournament. Winners move on to the next round. In essence; two robots enter, one robot leaves. (Walter, 2016, para. 5)

The possibility that the young adults described above would engage in dog or cock fighting is minimal. However, the battles in which they are engaging with robots can apparently elicit comparable levels of aggressive sentiment as well as vicarious aggression in onlookers.

Some of the issues involved in robot and autonomous vehicle abuse have close parallels with other "cyber-dehumanization" concerns such as violent video game utilization (Bastian et al., 2012), but also have increasingly complex and serious material dimensions. For example, many of the robots involved in abusive episodes play critical roles in everyday life, rather than narrative roles in video games. Robots along with other autonomous entities such as cars and trucks are being integrated into sensitive and essential transportation and manufacturing processes of organizations and communities. Many societies are entrusting their participants' health and wellbeing to them in critical settings (such as with autonomous vehicles on public roadways), and are often collaborating with them on various organizational initiatives, often without much consideration of security issues (Booth et al., 2017; Bragança et al., 2019; Oravec, 2017b; Rolenc, 2020; Winkelman et al., 2019). In contrast with some of the "machine wrecking" of the past, the autonomous entities that are being attacked are often delicate and intricate and the forms of damage inflicted can have comparable nuances; the robots can be reprogrammed or retrained to operate in dangerous ways that only emerge later, for example. The kinds of aggression described in this chapter often have a strong symbolic element and are somehow geared to destroy the robot's "dignity" (Hamilton & Mitchell, 2017) as well as functionality, reflecting the symbolic dimensions of human–robot encounters (Richardson, 2015). The destructibility (as subsequent reparability) of robots is indeed a benefit in many circumstances, as the robot's demise may provide an emotional outlet for individuals (as in the case of some sexbot usage). However, the consequences for such aggressive activities in terms of overall mental health are just beginning to surface (Danaher, 2017; Heath, 2016; Mackenzie, 2020). Various counter-trends are also evident that show the positive functions of displays of respect for robots: some robots that are considered "deceased" have been given funerals by their human workplace colleagues (Purtill, 2020), and robot deaths have often been characterized in serious and somber terms in social media (Carter et al., 2020).

Evolving Characterizations of Robots and Thinking Machines

The kinds of violence conducted against robots and autonomous entities can often be differentiated on the basis of their perceived levels of intelligence, with the intelligence being characterized in terms of science fiction stereotypes as well as engineering design. Robotics has evolved as a field as AI advances in machine learning have increased the autonomy with which robots can learn about their environments and make decisions. The International Federation of Robotics (IFR, 2019) defined a manufacturing robot as "an automatically controlled, reprogrammable and multipurpose manipulator for use in industrial automation applications, with "automation" concepts stemming from the work of Diebold (1952). The word "robot" is rooted in the 1920 Czech play (*Rossum's Universal Robots*) by Karel Čapek; the play became popular when performed in the US in the 1920s (Abnet, 2020). The term stems from the Slavic term "robota," for forced-labor worker, a name that signals the important role that robots play in workplace settings. Answers to the question of what is considered a robot have shifted as automation has become more sophisticated; as related by Burdick (1992), "Once a certain level of automation becomes widespread, we no longer call it robotic. For example, 200 years ago a dishwasher would have been considered a robot" (p. 2).

As robots take roles that are more tightly intertwined with human activities, characterizations of robots that are more socially-focused have emerged (Vanman & Kappas, 2019). Human mortality and vulnerability have often been factors in how robotics and AI applications have been construed in public policy and legal venues, with Isaac Asimov's (1950) "Three Laws of Robotics" an early attempt to provide some direction (Clarke, 1993; Dennett, 1997). Forms of "robophobia" and related "technophobia" have reportedly emerged (McClure, 2018; Paerregaard, 2019; Vanman & Kappas, 2019), some of which are linked to the "uncanny valley" phenomenon and its psychological associations with terror. Uncanny valley notions stem from empirical research showing that many individuals begin to recoil from robots as they become more similar, yet not quite duplicating, humans in appearance and behavior

(MacDorman, 2005; Mori, 2012). As described often in this book, the robot as an "other" has been a common theme of science fiction and various other creative works (Mayor, 2020). Coeckelbergh (2011) describes how the linguistic construction of artificial others (such as addressing them by name) affects the way individuals construe them as entities. The anthropomorphic dimensions of robots (such as a human-like face) are sometimes an explicit aspect of design and in other contexts are added features provided by users for entertainment (Corkery, 2020). The perceived levels and kinds of robot intelligence are also factors in the kinds of human–robot interactions that emerge (Banks, 2019).

The notion of autonomous "thinking machines" that can engage in functional and useful practices for workplaces as well as dysfunctional ones has permeated many robotics initiatives and AI approaches along with public consciousness concerning their applications for decades, framing some of the ethical considerations involving human-inflicted abuse issues. Such machines were generally construed as autonomous intelligent entities independently operating on the world with various degrees of autonomy and self-awareness (Abnet, 2020; Hurlburt, 2017; Oravec, 2019). Thinking machines have played roles in science fiction novels and other entertainment genres, often in the form of intelligent and mobile robots that can function either in positive or dysfunctional ways in various scenarios. Some scenarios of the future project that humans themselves will be displaced in their basic functions by intelligent systems (Heylighen & Lenartowicz, 2017), or become conjoined with robots in a "cyborg" configuration; other scenarios project that human beings will someday be able to upload their selves to these entities, in a form often construed as "artificial immortality" (Cave, 2020; DiCarlo, 2016; Ferrando, 2019). Humans have been frightened by such fictional robotic characters as *2001: A Space Odyssey's* HAL since the 1960s, and Mary Shelley's monster in *Frankenstein* played a comparable role in the 1800s (Shelley, 2012/1818). Research on how robots can express dominance in human–robot interaction has shown that various robot signals and positionings can increase human compliance (Peters et al., 2019). As discussed in this book, apparent superiority and control of humans by robots can be expressed in symbolic fashion as these autonomous entities are involved in judicial decision making (Sourdin, 2018), ostensibly to eliminate bias and reduce court expenditures.

Thinking machine notions can complicate the issue of workplace attacks on robots, introducing ethical issues beyond those of mere property destruction (Elhabashya et al., 2019; Whitby, 2008). For humans to attack an entity that is widely framed as "intelligent" presents complex factors of personal and social identity as well as of social and personal trust that go beyond attacks on mere "property." Trust in robots apparently also has some linkage to their levels of intelligence (Kim et al., 2020). Attempts to devalue robotic intelligence or desecrate robots can be attempts to transform the robotic entity being attacked to one that is less comparable to a human being and thus less deserving of non-aggressive, non-harassing treatment.

Varieties of Aggressive Behavior Involving Autonomous Entities

The machine wreckers in previous centuries had limited options for their attacks; however, many of those individuals were quite strategic in their activities, planning and timing their attacks for maximum impact (Linton, 1992). In the case of robot, AI, and autonomous vehicle abuse, the kinds of damage can include complex schemes to exploit the known weaknesses of the entities as well as more primitive attacks (Meryem & Mazri, 2019); often, narratives or images of the attacks are disseminated on social media (Taylor, 2022). The cleverness of the attacks, as well as the surprise with which they were planned, are often factors in attempts to place the humans involved in dominant positions over the robots or autonomous entities (Küster et al., 2021; Luria et al., 2020a).

Below are recent examples of aggressions toward robots and other AI-enhanced entities. They can differ in whether the entity attacked is considered in some ways as having human-style intelligence (robot as personage) or whether it is construed as primarily an object owned by a particular individual, organization, or community (robot as property):

Destruction of physical and decision-making capabilities: The kinds of autonomous entity destruction can vary if the damage is focused on the aspects of the entity that are seen as "intelligent"; attacks upon the robot's decision-making features can potentially cause less visible damage but can be as devastating in terms of eventual resource losses (Bartneck & Keijsers, 2020). For example, Halfacree (2021) describes emerging "poltergeist attacks" that could "leave autonomous vehicles blind to obstacles or haunt them with new ones." PenzeyMoog and

Slakoff (2021) relate "gaslighting" incidents in homes in which AI-enhanced "Internet of Things" technologies were surreptitiously manipulated in order to unsettle home residents. As robots and AI control structures are manipulated without attribution or apparent linkage to specific individuals and motives, some gaslighting patterns are emerging (Sweet, 2019). Robotic and autonomous entity destruction at the physical level can be rooted in the deficiencies or shortcomings of specific materials and structures used in automation (Ranabhat et al., 2019). Whether these complex factors are being considered in full by someone who is angry and strikes out at a specific robot is uncertain, however.

Image-related indignities: Some robots have been abused in ways that would signal to other humans visibly that some sort of desecration or humiliation of the entity has been conducted (as in Fraser et al., 2019; Harrison, 2019); sexually-themed indignities may be included here. If the robot is considered in terms of its own intelligence and personage the quality of this aggression can vary from that of an attack on the property of humans to an attack on an autonomous entity. The distribution of these images via social media (and related publicity) is often involved: for example, dogs have been placed in the "co-pilot" seats of moving Tesla self-driving vehicles as social media stunts, which has alarmed some animal welfare activists (Lambert, 2021).

Strategic or intentional neglect: Robots in dangerous or sensitive settings that are not monitored with prescribed levels of attention could cause significant damage to humans, themselves, and their environments. Intentional abuse of this kind is difficult to detect and those responsible identified until it produces major damage.

Verbal abuse: For humans to act out aggression in the form of verbal abuse and harassment toward robots and chatbots has become common (Neff & Nagy, 2016), with some of the harassment falling into the "bullying" category (D'Cruz & Noronha, 2021). Many chatbot developers are integrating ways to deal with sustained verbal abuse into their protocols (Chin & Yi, 2019). Much of this abuse has specific gender-related overtones (Strait et al., 2018; Taylor, 2022), raising issues involving the anthropomorphization of robots. Winkle et al. (2021) have proposed that robots that face such gender-themed abuse respond aggressively with comebacks that reflect a "feminist" perspective. They claim that through their proactive efforts "we were able to increase girls' perceptions of robot credibility and reduce gender bias in boys" (p. 29).

Manipulation and gaming, and the crowdsourcing of gaming strategies: For example, the "griefing" and bullying of autonomous vehicles in ways that attempt to confuse them or thwart their efforts has become a matter of concern for developers (Kiss, 2019; Liu et al., 2020; Moore et al., 2020); many griefing and manipulation strategies are being shared online.

Security breaches: The "hacking" of autonomous entities provides a general set of concerns for developers (Greenberg, 2017; Willison & Warkentin, 2013), with successful exploits being shared online and thus an escalation of hacking efforts and expertise. Such hacking can have as a by-product the loss of trust in robotics as well as increased resource investments. These robot hacking strategies can include ransomware and malware (Mayoral-Vilches et al., 2020).

Abuse rooted in subservient themes and positions: Slave- and subjugation-related notions have long roots in high technology and modern organizational life (Hampton, 2015; Ras & Gregoriou, 2019). Slavery has wrought centuries of pain in societies throughout the world. The notion that robots are often construed as workplace "slaves" may have increasingly problematic implications for ethical conduct in organizational life besides the occasional abusive events and related images that have already emerged.

As organizations and communities find that the introduction of autonomous entities into various settings is economically desirable, basic understandings about their utilization are being wrought that involve power considerations as well as ethical perspectives (Curchod et al., 2019). What kind of dignity is to be afforded these entities (and how their appropriate treatment is to be enforced) is among these agreements. Robots have strong symbolic meanings in workplace and community contexts, whether or not they are successfully integrated into various organizational service, industrial, or entertainment functions (Beane, 2020). The very presence of robots can apparently have an effect on some individuals who are engaging in various tasks, with apparent improvements in human performance (Spatola et al., 2018). Some insights as to how understandings may develop concerning robots could come from how dombots and sexbots are reportedly being treated. "Dombots," or domestic robots, are designed to relieve individuals of various routine household chores (McKnight, 2014), and sexually-themed robots ("sexbots") have been integrated into some practices

(Heath, 2016; Mackenzie, 2020). An assortment of unsettling physical and programming-related abuses of dombots and sexbots have been emerging that signal that aggression toward autonomous entities should be monitored for potential exacerbations of mental illness and criminal sexual offenses, especially involving children (Brown & Shelling, 2019).

Mitigation Efforts for Anti-Robot and Autonomous Entity Abuse

Many workplaces are being redesigned so that robots can function more efficiently, with the human being often construed as the alien in the robotic environment rather than the reverse (Miller, 1983). Manipulation and shaping of humans so that they fit in more adequately with the perceived requirements of workplace operations has had psychological as well as physical dimensions. Robots and automation as a whole are often promoted in workplaces as symbols of security and advancement (Beane, 2020) and even superiority to human workers, despite various technical deficiencies. However, forms of aggression toward robots are continuing, which can influence the character or human-to-human interactions as well (Carlson et al., 2019; Ringler & Reckter, 2012). Investigations into the matter of robot destruction have practical implications as organizational investments in robotics increase. Research on how to diminish the tendencies toward robot abuse has expanded considerably in the past few years. Monitoring levels of abuse of robots, chatbots, and other AI-enhanced entities can have apparent value for organizations in signaling disquiet and in otherwise observing employees and community members, which could be seen as an important source of surveillance in the near future (Schrage, 2016). Governmental and community regulations concerning the scope of activities of delivery, service, manufacturing, and related robots are often lacking (Trombola, 2021).

In many circumstances, the abuse of robots and other autonomous entities by humans has an emotional or expressive component, and countering this abuse can involve an emotional response from the attacked entity; first impressions of robots may affect these emotion-rooted interactions (Spence et al., 2014) as well as the emotional quality of the robot's responses to abuse (Connolly, 2020). Questions such as "Do a robot's social skills and its objection discourage interactants from switching the robot off?" (Horstmann et al., 2018, p. 1) are being asked and related research conducted in attempts to mitigate potential abuse. Murrer

(2020) describes how mobile delivery robots often have warning buzzers to ward off potentially abusive human contact. Lucas et al. (2016) show that robot size may be a factor in the kind and quality of abuse, and some robots may be perceived to be "too big to abuse." Encouraging human bystanders to intervene when there is robotic abuse is another mitigation approach (Tan et al., 2018), just as in human-to-human bullying where bystanders can be an ameliorating factor. Protecting individuals from robots with particular protective clothing and shielding may alter their characterizations of the robots they work with, potentially making the humans feel more impervious and less threatened. For example, Palmer (2019) describes how Amazon Corporation's "anti-robot vest" is designed to "protect workers from giant machines smashing into them" with the strategic placement of various sensors. Anti-robot warnings and precautions may be warranted if the robots have potentially been manipulated or tampered; some robots may indeed be "rogue," with operations hacked in ways that benefit external competitors or saboteurs (Maggi et al., 2017; Reynolds & Ishikawa, 2006; Wolfert et al., 2020).

Some organizations may take activist approaches to the phenomena associated with robot abuse, providing settings in which employees can openly express their antagonism to autonomous entities with little overall damage to the organization. Such expressions of aggression toward robots could indeed identify and expose individuals who have overall problems with aggression in organizational contexts (Sparrow, 2020), or who have specific grievances. Approaches for designing robots that are relatively impervious to abuse or easily reconstructable have been proposed (Cohen, 2019; Luria et al., 2020a). Strategies in which autonomous entities can be programmed to "feel" and express a kind of "pain" or "punishment" can be part of such approaches (Keijsers et al., 2019; Richardson et al., 2020; Rossmy et al., 2020; Sandberg, 2015), providing some means for individuals to express their aggression toward robots without physically damaging them or human bystanders. Strategies for programming pleasure and related motivating factors have also been proposed (Lewis & Canamero, 2016), giving individuals means of applying forms of behavioral modification rather than punishment in their interactions with robots.

Many anthropomorphism initiatives intended to ameliorate robots to workers and community participants could be misplaced, with anti-robot antagonism being intensified with the presence of an "uncanny valley" entity, as previously discussed (Mori, 2012). Solutions to robot abuse problems that are rooted in making robots seem "nicer" are proliferating (Cappuccio et al., 2019), but need to be reconsidered in light of the complex social and economic factors involved. Individuals use aggression to express their anxieties and fears pertaining to their economic or social statuses, whether they characterize the proximate object of their aggression as pleasant or disagreeable. Research by Küster et al. (2021) using short videos of robots being abused shows how some humans can "be responsive to harm done to robots—while simultaneously feeling discomforted and concerned" (p. 3327), potentially showing the way for individuals to become less openly aggressive toward robots in workplace and community contexts.

Dystopian Futures for Robots and AI Entities: Demonizing and Weaponizing

Scenarios in which individuals will ultimately need to engage in battles with AI-enhanced robot "overlords" have been discussed by such luminaries as Elon Musk and Stephen Hawking (Leitao, 2019; Thompson, 2018), with a coming "robot war" often projected (Singer, 2009). Many acts of destruction by humans are apparently driven by anxieties and fears related to technological changes, so the continuance and even expansion of attacks on robots have been projected by some researchers (Bartneck & Keijsers, 2020; Bigliardi, 2019). Robots are being utilized strategically in many aggressive criminal operations, including their weaponization by using them in physical or virtual security breaches (de Ágreda, 2020), so defensive approaches toward robots and other AI-related entities are expanding on a number of dimensions. Sexually-themed robots can provide some insights as to how robots can be used for dysfunctional and abusive purposes as well as more moderate and even life-enhancing functions (Heath, 2016; Mackenzie, 2020; McKnight, 2014). The dimensions of the human imagination that support violence, anger, and aggression apparently have a great potential for realization in robots and autonomous entities (Lewnard, 2020; Reyes et al., 2019). Such phenomena as the organized hunting of robots by humans could indeed follow; robots are already being taught to hunt other robots using pack hunting strategies (Grossman, 2016).

Robots that are specially designed to elicit certain human emotions may be used to mitigate some trends toward anti-robot aggression (as in Shao et al., 2020; Vinanzi et al., 2019). For instance, Briggs and Scheutz

(2014) explored the effects of various robotic displays of protest and distress on the humans with whom the robots are situated, and Reyes, Meza, and Pineda analyzed assorted displays of anger. Robotic exterior appearances have been shown in some cases to "solicit compassion and attachment in humans, and [their] cognitive resources may be powerful enough to establish enduring and relatively rich relationships with their users" (Cappuccio et al., 2019, p. 10). However, if these strategies fail to mitigate anti-robot aggression, managers may choose to accept and even capitalize on a certain level of robot abuse, with robot designers building in the capacity to endure focused human abuse. Attracting some amount of abuse to robots may keep individuals from engaging in attacks that could directly hurt human beings, deflecting particular aggressions. The strategic use of anti-robot violence for cathartic and expressive purposes in workplace contexts projects some potentials, just as video game interactions with humans fighting robotic figures can apparently provide catharsis for some individuals (Bastian et al., 2012), though with some possible side effects in terms of their perceptions of the acceptability of violence.

How robots are treated in particular contexts may soon help to shape how mental activity and intelligent control functions are construed, themes of special interest and concern with the proliferation of cyborgs. For example, human–machine hybrids are emerging as the availability of cybernetic implants expands (Egan, 2020); some of the antagonism toward robots described in this chapter may unfortunately extend to these cyborgs. The extent to which organizations allow for the open destruction of robots may eventually affect how humans (considered as a related form of intelligent entity) perceive and interact with each other (Coghlan et al., 2019). Such rituals as the previously-described robot "funerals" may make robots appear to have increased value within the organization. Even though the entities involved are indeed machinery, the approaches organizations take toward their treatment can affect the character of the organization as a whole.

Decades ago, Stern (1937) characterized various "resistances to the adoption of technological innovations," with "machine wreckers" (Hodson, 1995) expressing dissatisfaction with automation. Destructions or sabotages of robots by organizational participants (whether or not they are related to such deaths) could also produce substantial economic losses, and perhaps even precipitate collateral injury on the part of workers. Currently, the potentials that many individuals will smash robots and destroy autonomous entities are relatively small; however, as the use of robots in workplaces continues, and economic and social anxieties compound the situation, the possibility for series of acts of sabotage and possibly more substantial damages could increase. Legal solutions are often limited for individuals who are injured by robots because of the diffuse responsibility for these occurrences; multiple levels of designers, developers, and testers as well as human operators are potentially at legal fault for the operations of an autonomous entity (Prianto, et al., 2020).

Some Conclusions and Reflections

Why would someone attack a robot or autonomous vehicle? Such abuses have indeed become sport in some arenas, as expressed about the US television show *BattleBots* on a sports-themed website: "Who the hell even thought of constructing not only crazy robots with different types of weapons on them but also an arena with insane destructive capabilities. This is the same type of attitude that got us the coliseum with the human sacrifices" (Football, 2021). The abuse of robots and other autonomous entities can indeed reflect little more than momentary, situational anger, aggression, and anxieties. However, more serious social and psychological phenomena are often involved. Forms of cognitive and psychological overload concerning robots and AI can affect individuals negatively, from the mental demands of monitoring these entities' movements to the economic stresses involved with one's own possible replacement. For example, expressive dimensions of intelligent entity abuse can involve linkages to larger concerns about automation as well as various economic struggles (Joosse, 2020), aspects that are important to recognize as organizations and communities endeavor to prevent anti-robot attacks and mitigate the related damages.

As outlined throughout this book, "robot versus human" narratives are pervasive in many societies (manifest in science fiction as well as news and economic reports), so connections in everyday life with anti-robot themes are readily made; the contests and demonstrations that supposedly establish that robots "outclass" humans provide a major example here (as discussed in previous chapters). Even without more formalized varieties of human resistance, such as formations and proliferations of organized protest or insurrection, an anti-robot perspective can permeate societal discourse. Young people are presented with streams of negative images of robot activities through movies, comics, anime, and other genres.

Disseminating images and accounts of robot abuse through broadcast or social media can amplify the impacts of any particular anti-robot incident perpetrated by humans (Mrug et al., 2008), often expanding the related anti-automation social connections and discourse. Fictionalized narratives portraying employees who act out against the frustrations of technology by enacting violent behaviors against computer technology also proliferate (Alexander & Kochheiser, 2017). Autonomous vehicles are increasingly becoming the objects of abuse, with vehicles that violate individuals' expectations especially targeted (Smith et al., 2020). Robots, autonomous vehicles, and other autonomous entities can certainly be "gamed" in ways that may not seem to be overly abusive to those involved, with the hacking and manipulation of these devices construed as pleasurable by their perpetrators. As discussed in the chapter on sex robots, having certain forms of intimate interaction with robots may continue this theme of aggressive and even abusive anti-robot activity.

Supporting classes of physically-fragile robotic and vehicular entities that can be abused and manipulated by humans, often without detection, can be resource intensive for organizations; various legal consequences may be involved as well as specific physical damages (Lemley & Casey, 2019). Humans may try to attack robots by altering the robots' environments as well, making them less able to function as designed. Various metaverse environments and connections may also be altered in ways that jeopardize the functioning of the robots. For the past decades, humans have been working in environments that are increasingly arranged for the needs of robots rather than humans (Kisska-Schulze & Davis-Nozemack, 2020; Lafontaine, 2020; Miller, 1983). In some of those settings, autonomous entities are placed in positions that are seemingly superior than those assigned to human individuals (Haenlein & Kaplan, 2019; Leitao, 2019), possibly setting up conditions in which humans would want to respond and express themselves in some aggressive or otherwise negative manner. Robot and autonomous entity abuse and sabotage as social practices have expanded individuals' ranges of expression of their attitudes toward automation and other societal changes, although they have also had unfortunate consequences for human safety and organizational resources. For example, humans attacking robots can do so in ways that are aimed to injure the robot directly and "personally" (in ways comparable to delivering an attack upon another human), or that are primarily aimed to do damage to the organization as a whole (e.g., by requiring it to make repairs to the robot or replace it). Robots are often fragile and brittle in some aspects, and considerable robot abuse can indeed be "casual" breakage; a robot or autonomous vehicle can simply be in the way of the person or persons perpetrating a particular act of violence. However, the potentials for more intentional or planned attacks (such as the malicious reprogramming or systematic diversion of robots, or significant alterations of their environments) present unsettling prospects for human safety, especially in an era when the use of autonomous vehicles is expanding. Many workplaces and community centers present confined settings in which individuals' expressions of violence and bullying can have significance for more individuals than just the perpetrator and target, and the audiences for the violence can be affected by fearing or by applauding the perpetrators (Graver, 1995).

Although many narratives of robot abuse have widespread dissemination through social media, much of what happens in workplaces and community centers in terms of sabotage is still not readily accessible outside of the organization, with few employers and administrators wanting to publicize such occurrences (Singh, 2020). When accounts of abuse do surface, however, there is often widespread discourse in social media and other popular forums on the social and ethical dimensions of the relationships between humans and robots (e.g., Cha et al., 2020; Corkery, 2020; Fraser et al., 2019; Harrison, 2019; Romero, 2018). Workplaces and community settings provide various artifacts and contexts for rituals and routines that can impart social and cultural information about superiority, subservience, and other aspects of the interaction order (Goffman, 1983), an order that increasingly involves robots and other AI-enhanced entities. The intense involvement of autonomous entities in the construction of this social realm has widened the scope of what organizational participants can manifest in terms of their positive sentiment and appreciation and also of aggression and abuse.

Individuals are facing an unsettling amalgam of problematic working and community concerns, rooted in perceived economic competition and related tensions. The strategy of many designers to make robots and self-driving vehicles more humanlike in order to ameliorate potential tensions about automation may not mitigate aggression against autonomous entities, given these circumstances. Robots that are anthropomorphized provide opportunities for abusive conduct that has close parallels to human abuse and bullying rituals and can unfortunately extend and enhance the patterns involved, as well as potentially create new ones. The dehumanization and devaluation of robots can serve as a mechanism for moral disengagement, in comparable manners to human-to-human workplace bullying (Smokowski & Evans, 2019). Such abuse can consume resources and distract from larger organizational undertakings, as well as potentially result in the injuries or deaths of human bystanders (Maggi et al., 2017; Sanderson et al., 1986). In some limited settings, the strategy for dealing with the abuse of autonomous entities may be to consider it as basically harmless activity (except for physical damage), and as potentially useful in siphoning the energies that would otherwise be directed toward human bullying or other attacks (Carlson et al., 2017). In many contexts, robots could become the kinds of "punching bags" that can indeed deflect and lessen open expressions of aggression toward other humans in certain settings (McLinton et al., 2018) as well as have a cathartic effect (Luria et al., 2020a). Robot anthropomorphism may be usefully extended to include the potentials for attracting and receiving abuse in ways that provide workers with catharsis or other social expressions, or of repelling and dissipating aggression (Cohen, 2019).

The construction of a robotic "other" in workplace or community contexts does not start from scratch; most individuals have gleaned from an early age substantial cultural insights about robotics through popular writings, film, and news (Higbie, 2013; Paerregaard, 2019; Teo, 2020). The overall influence on workplaces of all of these social reflections about and images of robots is still uncertain, especially given the changes in robotics (which include cyborg-style implantations of robotic parts in human prostheses). Despite the growing uncertainties involved with robot operations and security (as described in Ionescu, Schlund, & Schmidbauer, 2019), robots are increasingly portrayed by administrators and developers as non-problematic companions and collaborators (Abnet, 2020; Melson et al., 2009). The objective of framing robots so as to be friendly colleagues in the confined settings of the workplace may backfire as they become unstable and untrustworthy targets of sabotage and abuse (Kinzler et al., 2019). Construing robots as supervisors and educators (and as somehow superior to humans) can indeed unsettle the situation even further (Sahota & Ashley, 2019); such forms of symbolic violence against individuals may be countered with human-generated abuse and manipulation.

The potentially positive dimensions of the robots-versus-humans formulation should certainly not be overlooked. Discourse on robotic "others" could indeed draw humans together by emphasizing their common humanity in contrast to their robotic colleagues (Coghlan et al.,

2019; Jackson et al., 2020); Spatola (2020) explores how the comparison of robots with humans on essential dimensions such as mortality can unite humans. However, in this position of "otherness," robots and other autonomous entities could also be more vulnerable to attack, which can ultimately present dangers to human beings. Science fiction pioneer Ray Bradbury's proactive approach may be useful here in inspiring mitigation efforts: "I don't try to describe the future. I try to prevent it" (quoted by Arthur C. Clarke, 1992). Organizational and community participants can work with technology developers to understand the kinds of economic and symbolic violence that are linked (however inappropriately) with autonomous entities and help to counter their abuse by humans with these insights.

References

Aaltonen, I., Arvola, A., Heikkilä, P., & Lammi, H. (2017, March). Hello Pepper, may I tickle you? Children's and adults' responses to an entertainment robot at a shopping mall. In *Proceedings of the Companion of the 2017 ACM/IEEE International Conference on Human-Robot Interaction* (pp. 53–54).

Abnet, D. A. (2020). *The American robot: A cultural history*. University of Chicago Press.

Alexander, R., & Kochheiser, S. (2017). Office Space as hyperreality—Using film as a postmodern critique of bureaucracy. *Public Voices, 10*(1), 80–98.

Alexis, K. (2020). Towards a science of resilient robotic autonomy. Cornell University, *arXiv preprint* arXiv:2004.02403

Asimov, I. (1950). *I, Robot*. Bantham Books.

Bacon, A. M., May, J., & Charlesford, J. J. (2021). Understanding public attitudes to hate: Developing and testing a UK version of the Hate Crime Beliefs Scale. *Journal of Interpersonal Violence, 36*(23–24), NP13365–NP13390. https://doi.org/10.1177/0886260520906188

Bankins, S., & Formosa, P. (2020). When AI meets PC: Exploring the implications of workplace social robots and a human-robot psychological contract. *European Journal of Work and Organizational Psychology, 29*(2), 215–229.

Banks, J. (2019). Theory of mind in social robots: Replication of five established human tests. *International Journal of Social Robotics, 12*, 403–414. https://doi.org/10.1007/s12369-019-00588-x

Bartneck, C., & Keijsers, M. (2020). The morality of abusing a robot. *Paladyn, Journal of Behavioral Robotics, 11*(1), 271–283.

Bastian, B., Jetten, J., & Radke, H. R. (2012). Cyber-dehumanization: Violent video game play diminishes our humanity. *Journal of Experimental Social Psychology, 48*(2), 486–491.

Beane, M. (2020, March). In storage, yet on display: An empirical investigation of robots' value as social signals. In *Proceedings of the 2020 ACM/IEEE International Conference on Human-Robot Interaction* (pp. 83–91).

Bernal, N. (2019, September 29). British workers are deliberately sabotaging robots amid fears they will take their jobs, study finds. *Telegraph*. https://www.telegraph.co.uk/technology/2019/09/29/british-workers-deliberately-sabotaging-robots-amid-fears-will/

Bergek, A., Berggren, C., Magnusson, T., & Hobday, M. (2013). Technological discontinuities and the challenge for incumbent firms: Destruction, disruption or creative accumulation? *Research Policy, 42*(6–7), 1210–1224.

Bernhardtz, V. E. (2019). Black boxes of cognitive computers and the impact on labor markets. In A. Larsson & R. Teigland (Eds.), *The digital transformation of labor* (pp. 100–117). Routledge.

Bharadwaj, A., & Dvorkin, M. A. (2019). The rise of automation: How robots may impact the U.S. labor market. *The Regional Economist* (Federal Reserve Bank of St. Louis), *27*(2). https://ideas.repec.org/a/fip/fedlre/00220.html

Bigliardi, S. (2019). "We don't know exactly how they work": Making sense of technophobia in 1973 *Westworld*, *Futureworld*, and *Beyond Westworld*. *Journal of Science Fiction and Philosophy, 2*.

Black, D. (2019). Machines with faces: Robot bodies and the problem of cruelty. *Body & Society, 25*(2), 3–27.

Booth, S., Tompkin, J., Pfister, H., Waldo, J., Gajos, K., & Nagpal, R. (2017, March). Piggybacking robots: Human-robot overtrust in university dormitory security. In *Proceedings of the 2017 ACM/IEEE International Conference on Human-Robot Interaction* (pp. 426–434).

Bourdieu, P. (1991). *Language and symbolic power*. Harvard University Press.

Bragança, S., Costa, E., Castellucci, I., & Arezes, P. M. (2019). A brief overview of the use of collaborative robots in industry 4.0: Human role and safety. In *Occupational and environmental safety and health* (pp. 641–650). Springer.

Briggs, G., & Scheutz, M. (2014). How robots can affect human behavior: Investigating the effects of robotic displays of protest and distress. *International Journal of Social Robotics, 6*(3), 343–355.

Brown, R., & Shelling, J. (2019). Exploring the implications of child sex dolls. *Trends and Issues in Crime and Criminal Justice* (570), 1–13. https://search.informit.com.au/documentSummary;dn=332641507717973;res=IELAPA

Burdick, J. W. (1992). Robots that crawl, walk, and slither. *Engineering & Science, 55*(4), 2–13.

Cappuccio, M. L., Peeters, A., & McDonald, W. (2019). Sympathy for Dolores: Moral consideration for robots based on virtue and recognition. *Philosophy & Technology, 30*, 9–31. https://doi.org/10.1007/s13347-019-0341-y

Carlson, Z., Lemmon, L., Higgins, M., Frank, D., & Feil-Seifer, D. (2017). This robot stinks! Differences between perceived mistreatment of robot and computer partners. *arXiv preprint* arXiv:1711.00561

Carlson, Z., Lemmon, L., Higgins, M., Frank, D., Shahrezaie, R. S., & Feil-Seifer, D. (2019). Perceived mistreatment and emotional capability following aggressive treatment of robots and computers. *International Journal of Social Robotics, 11*(5), 727–739.

Carter, E. J., Reig, S., Tan, X. Z., Laput, G., Rosenthal, S., & Steinfeld, A. (2020, March). Death of a robot: Social media reactions and language usage when a robot stops operating. In *Proceedings of the 2020 ACM/IEEE International Conference on Human-Robot Interaction* (pp. 589–597).

Cauchard, J. R., Zhai, K. Y., Spadafora, M., & Landay, J. A. (2016, March). Emotion encoding in human-drone interaction. In *2016 11th ACM/IEEE International Conference on Human-Robot Interaction (HRI)* (pp. 263–270). New York: IEEE Press.

Cave, S. (2020). AI: Artificial immortality and narratives of mind uploading. *AI Narratives: A History of Imaginative Thinking about Intelligent Machines*, 309.

Cha, Y. J., Baek, S., Ahn, G., Lee, H., Lee, B., Shin, J. E., & Jang, D. (2020). Compensating for the loss of human distinctiveness: The use of social creativity under Human-Machine comparisons. *Computers in Human Behavior, 103*, 80–90.

Chapman, C. (2017). Computer Says "Ouch." *Daily Mail.* https://www.dailymail.co.uk/news/article-4895882/Robot-gets-beaten-baseball-bat-Moscow.html

Chin, H., & Yi, M. Y. (2019, May). Should an agent be ignoring it? A study of verbal abuse types and conversational agents' response styles. In *Extended Abstracts of the 2019 CHI Conference on Human Factors in Computing Systems* (pp. 1–6).

Clarke, A. C. (1992, July 16). Interview with Ray Bradbury. *Independent*.

Clarke, R. (1993). Asimov's laws of robotics: Implications for information technology-Part I. *Computer, 26*(12), 53–61.

Coeckelbergh, M. (2011). You, robot: On the linguistic construction of artificial others. *AI & Society, 26*(1), 61–69.

Coghlan, S., Vetere, F., Waycott, J., & Neves, B. B. (2019). Could social robots make us kinder or crueller to humans and animals? *International Journal of Social Robotics, 11*(5), 741–751.

Cohen, N. (2019). Making the case for robotic objects as anger outlets. *Techxplore*, https://techxplore.com/news/2019-05-case-robotic-anger-outlets.html

Condliffe, J. (2016, November 3). Humans will bully mild-mannered autonomous cars. *Technology Review*. https://www.technologyreview.com/2016/11/03/156270/humans-will-bully-mild-mannered-autonomous-cars/

Connolly, J. (2020, March). Preventing robot abuse through emotional robot responses. In *Companion of the 2020 ACM/IEEE International Conference on Human-Robot Interaction* (pp. 558–560).

Connolly, J., Mocz, V., Salomons, N., Valdez, J., Tsoi, N., Scassellati, B., & Vázquez, M. (2020, March). Prompting prosocial human interventions in response to robot mistreatment. In *Proceedings of the 2020 ACM/IEEE International Conference on Human-Robot Interaction* (pp. 211–220).

Corkery, M. (2020, February 26). Should robots have a face? *New York Times*. https://www.nytimes.com/2020/02/26/business/robots-retail-jobs.html

Cox, K. (2020, February 27). Opinion: Starship delivery robots won't benefit campus. *The Breeze*. https://www.breezejmu.org/opinion/opinion-starship-delivery-robots-won-t-benefit-campus/article_6027969e-58cc-11ea-a4a3-eb6820173e04.html

Crowley, M. (2014). Class, control, and relational indignity: Labor process foundations for workplace humiliation, conflict, and shame. *American Behavioral Scientist, 58*(3), 416–434.

Culver, G. (2018). Death and the car: On (auto) mobility, violence, and injustice. *ACME: An International Journal for Critical Geographies, 17*(1), 144–170.

Curchod, C., Patriotta, G., Cohen, L., & Neysen, N. (2019). Working for an algorithm: Power asymmetries and agency in online work settings. *Administrative Science Quarterly, 65*(3), 644–676. https://doi.org/10.1177/0001839219867024

Czarniawska, B., & Joerges, B. (2020). *Robotization of work?: Answers from popular culture, media and social sciences*. Edward Elgar Publishing.

D'Cruz, P., & Noronha, E. (2021). Workplace bullying in the context of robotization: Contemplating the future of the field. *Concepts, Approaches and Methods*, 293–321.

Danaher, J. (2017). Robotic rape and robotic child sexual abuse: Should they be criminalised? *Criminal Law and Philosophy: An International Journal for Philosophy of Crime, Criminal Law and Punishment, 11*(1), 71–95.

de Ágreda, Á. G. (2020). Ethics of autonomous weapons systems and its applicability to any AI systems. *Telecommunications Policy, 44*(6). https://doi.org/10.1016/j.telpol.2020.101953

Dennett, D. (1997). Did HAL commit murder? In D. G. Stork (Ed.), *HAL's legacy: 2001's computer as dream and reality*. MIT Press.

DiCarlo, C. (2016). How to avoid a robotic Apocalypse: A consideration on the future developments of AI, emergent consciousness, and the Frankenstein effect. *IEEE Technology and Society Magazine, 35*(4), 56–61.

Diebold, J. (1952). *Automation: The advent of the automatic factory*. Van Nostrand.

DiFruscia, K. T. (2012). Work rage: The invention of a human resource management anti-conflictual fable. *Anthropology of Work Review, 33*(2), 89–100.

Dorson, R. M. (1965). The career of "John Henry." *Western Folklore, 24*(3), 155–163.

Egan, M. (2020, September 18). Workers fear humans implanted with microchips will steal their jobs. *CNN Business*. https://www.cnn.com/2020/09/18/business/jobs-robots-microchips-cyborg/index.html

Elhabashya, A. E., Wellsb, L. J., & Camelioc, J. A. (2019). Cyber-physical security research efforts in manufacturing–A literature. *Procedia Manufacturing, 34*, 921–931.

Eliot, L. (2019, June 4). Human drivers bullying self-driving cars: Unlawful or fair game? *Forbes*. https://www.forbes.com/sites/lanceeliot/2019/06/04/human-drivers-bullying-self-driving-cars-unlawful-or-fair-game/#1c6895ff49ee

Farivar, C. (2021, June 27). Security robots expand across U.S., with few tangible results. *NBCNews.com*. https://www.nbcnews.com/business/business-news/security-robots-expand-across-u-s-few-tangible-results-n1272421

Ferrando, F. (2019). The posthuman divine: When robots can be enlightened. *Sophia, 58*(4), 645–651.

Football, B. (2021, October 26). Robot fighting has gone too far. *Barstool Sports*. https://www.barstoolsports.com/blog/3390802/video-robot-fighting-has-gone-too-far

Fraser, K. C., Zeller, F., Smith, D. H., Mohammad, S., & Rudzicz, F. (2019, June). How do we feel when a robot dies? Emotions expressed on Twitter before and after hitchBOT's destruction. In *Proceedings of the Tenth Workshop on Computational Approaches to Subjectivity, Sentiment and Social Media Analysis* (pp. 62–71).

Froomkin, A. M., & Colangelo, P. Z. (2015). Self-defense against robots and drones. *Connecticut Law Review, 48*, 1.

Gnambs, T., & Appel, M. (2019). Are robots becoming unpopular? Changes in attitudes towards autonomous robotic systems in Europe. *Computers in Human Behavior, 93*, 53–61.

Goffman, E. (1983). The interaction order: American Sociological Association, 1982 presidential address. *American Sociological Review, 48*(1), 1–17.

Gorgan, E. (2019, September 4). Brits are sabotaging workplace robots out of fear they might lose their job. *Auto Evolution*. https://www.autoevolution.com/news/brits-are-sabotaging-workplace-robots-out-of-fear-they-might-lose-their-job-137933.html

Graver, D. (1995). Violent theatricality: Displayed enactments of aggression and pain. *Theatre Journal, 47*(1), 43–64.

Greenberg, A. (2017). Watch hackers hijack three robots for spying and sabotage. *Wired*. https://www.wired.com/story/watch-robot-hacks-spy-sabotage/

Grint, K., & Case, P. (1998). The violent rhetoric of re-engineering: Management consultancy on the offensive. *Journal of Management Studies, 35*(5), 557–577.

Grossman, D. (2016). Scientists are teaching robots how to hunt other robots. *Popular Mechanics*. https://www.popularmechanics.com/technology/robots/a21728/robots-hunting-other-robots-now/

Haenlein, M., & Kaplan, A. (2019). A brief history of artificial intelligence: On the past, present, and future of artificial intelligence. *California Management Review, 61*(4), 5–14.

Halfacree, G. (2021, June 18). Poltergeist attack could leave autonomous vehicles blind to obstacles – or haunt them with new ones. *The Register*. https://www.theregister.com/2021/06/18/poltergeist_autonomous_vehicles/

Hamilton, I. (2018, June 9). People kicking these food delivery robots is an early insight into how cruel humans could be to robots. *Business Insider*. https://www.businessinsider.com/people-are-kicking-starship-technologies-food-delivery-robots-2018-6

Hamilton, I. (2019, June 13). Uber says people are bullying its self-driving cars with rude gestures and road rage. *Business Insider*. https://www.businessinsider.com/uber-people-bullying-self-driving-cars-2019-6

Hamilton, L., & Mitchell, L. (2017). Dignity and species difference within organizations. In *Dignity and the Organization* (pp. 59–80). Palgrave Macmillan.

Hampton, G. J. (2015). *Imagining slaves and robots in literature, film, and popular culture: Reinventing yesterday's slave with tomorrow's robot*. Lexington Books.

Harrison, S. (2019, August 29). Of course citizens should be allowed to kick robots. *Wired*. https://www.wired.com/story/citizens-should-be-allowed-to-kick-robots/

Heath, H. (2016). Using/abusing fembots: The ethics of sex with robots. *Overland, 225*, 70.

Heylighen, F., & Lenartowicz, M. (2017). The Global Brain as a model of the future information society. *Technological Forecasting and Social Change, 114*, 1–6.

Higbie, T. (2013). Why do robots rebel? The labor history of a cultural icon. *Labor: Studies in Working-Class History of the Americas, 10*(1), 99–121.

Hodson, R. (1995). Worker resistance: An underdeveloped concept in the sociology of work. *Economic and Industrial Democracy, 16*(1), 79–110.

Hong, S. (2004). Man and machine in the 1960s. *Techné: Research in Philosophy and Technology, 7*(3), 50–78.

Horstmann, A. C., Bock, N., Linhuber, E., Szczuka, J. M., Straßmann, C., & Krämer, N. C. (2018). Do a robot's social skills and its objection discourage interactants from switching the robot off? *PLOS One, 13*(7), https://doi.org/10.1371/journal.pone.0201581

Hurlburt, G. (2017). Superintelligence: Myth or pressing reality? *IT Professional, 19*(1), 6–11.

Hyuga, T., & Alpeyev, P. (2021, June 29). SoftBank halts production of $1,800 Pepper humanoid robot. *Bloomberg.com.* https://www.bloomberg.com/news/articles/2021-06-29/softbank-mothballs-once-hyped-1-800-pepper-humanoid-robot

IFR, International Federation of Robotics. (2019). https://ifr.org/#topics. Accessed June 29, 2020.

Ionescu, T. B., Schlund, S., & Schmidbauer, C. (2019, July). Epistemic debt: A concept and measure of technical ignorance in smart manufacturing. In *International Conference on Applied Human Factors and Ergonomics* (pp. 81–93). Springer.

Jackson, J. C., Castelo, N., & Gray, K. (2020). Could a rising robot workforce make humans less prejudiced? *American Psychologist.* https://psycnet.apa.org/doi/10.1037/amp0000582

Jones, S. E. (2013). *Against technology: From the Luddites to neo-Luddism.* Routledge.

Joosse, P. (2020). Narratives of rebellion. *European Journal of Criminology.* https://doi.org/10.1177/1477370819874426.

Keijsers, M., & Bartneck, C. (2018, February). Mindless robots get bullied. In *Proceedings of the 2018 ACM/IEEE International Conference on Human-Robot Interaction* (pp. 205–214).

Keijsers, M., Kazmi, H., Eyssel, F., & Bartneck, C. (2019). Teaching robots a lesson: Determinants of robot punishment. *International Journal of Social Robotics*, 1–14. https://doi.org/10.1007/s12369-019-00608-w

Kim, W., Kim, N., Lyons, J. B., & Nam, C. S. (2020). Factors affecting trust in high-vulnerability human-robot interaction contexts: A structural equation modelling approach. *Applied Ergonomics, 85.* https://doi.org/10.1016/j.apergo.2020.103056

Kinzler, M., Miller, J., Wu, Z., Williams, A., & Perouli, D. (2019, April). Cybersecurity vulnerabilities in two artificially intelligent humanoids on the market. In *Workshop on Technology and Consumer Protection (ConPro '19), held in conjunction with the 40th IEEE Symposium on Security and Privacy.* https://par.nsf.gov/biblio/10099177

Kiss, G. (2019, August). External manipulation of autonomous vehicles. In *2019 IEEE SmartWorld, Ubiquitous Intelligence & Computing, Advanced & Trusted Computing, Scalable Computing & Communications, Cloud & Big Data Computing, Internet of People and Smart City Innovation (SmartWorld/SCALCOM/UIC/ATC/CBDCom/IOP/SCI)* (pp. 248–252). IEEE Press.

Kisska-Schulze, K., & Davis-Nozemack, K. (2020). Humans vs. Robots: Rethinking policy for a more sustainable future. *Maryland Law Review, 79*(4), 19–107.

Kolodny, L. (2018, June 18). Elon Musk emails employees about 'extensive and damaging sabotage' by employee. *CNBC.* https://www.cnbc.com/2018/06/18/elon-musk-email-employee-conducted-extensive-and-damaging-sabotage.html

Kracher, B. (2005, July). 21st century protests against objectionable labor practices. In *Proceedings of the International Association for Business and Society* (Vol. 16, pp. 326–329).

Krumins, A. (2017, May 4). Why the surge in violence against robots matters. *Extreme Tech.* https://www.extremetech.com/extreme/248806-surge-violence-robots-matters

Küster, D., Swiderska, A., & Gunkel, D. (2021). I saw it on YouTube! How online videos shape perceptions of mind, morality, and fears about robots. *New Media & Society, 23*(11), 3312–3331.

Lafontaine, C. (2020, July). Towards lively surveillance? The domestication of companion robots. In *International Conference on Human-Computer Interaction* (pp. 486–496). Springer.

Lambert, F. (2021, December 14). Idiots use Tesla Autopilot to put dog in danger in attempt to go viral. *Electrek.* https://electrek.co/2021/12/14/idiots-use-tesla-autopilot-put-dog-in-danger-attempt-viral/

Lee, J. D. (2020). Driver trust in autonomous, connected, and intelligent vehicles. In D. Fischer (Ed.), *Handbook of human factors for autonomous, connected, and intelligent vehicles.* CRC Press.

Leitao, M. (2019). *All hail our robot overlords: Obedience to digital authority* (Doctoral dissertation, California State University, Northridge).

Lemley, M. A., & Casey, B. (2019). Remedies for robots. *The University of Chicago Law Review, 86*(5), 1311–1396.

LeVasseur, T. (2017). Decisive ecological warfare: Triggering industrial collapse via deep Green resistance. *Journal for the Study of Religion, Nature & Culture, 11*(1), 109–130.

Levin, T. (2021). A man arrested for riding in the back seat of his driverless Tesla got out of jail, bought a new one, and did it again. *Business Insider.* https://www.businessinsider.com/tesla-fsd-back-seat-driving-stunt-arrested-buys-new-car-2021-5

Lewis, M., & Canamero, L. (2016). Hedonic quality or reward? A study of basic pleasure in homeostasis and decision making of a motivated autonomous robot. *Adaptive Behavior, 24*(5), 267–291.

Lewnard, J. (2020, February 28). *Robot Rumble features student machines ready for destruction*. https://www.dailyherald.com/news/20200228/robot-rumble-features-student-machines-built-for-destruction

Linton, D. (1992). The Luddites: How did they get that bad reputation? *Labor History, 33*(4), 529–537.

Liu, P., Du, Y., Wang, L., & Da Young, J. (2020). Ready to bully autonomous vehicles on public roads? *Accident Analysis & Prevention, 137*, 105457.

Lloyd, M. (2017). On the way to cycle rage: Disputed mobile formations. *Mobilities, 12*(3), 384–404.

Lloyd, C., & Payne, J. (2019). Rethinking country effects: Robotics, AI and work futures in Norway and the UK. *New Technology, Work and Employment, 34*(3), 208–225.

Lucas, H., Poston, J., Yocum, N., Carlson, Z., & Feil-Seifer, D. (2016, August). Too big to be mistreated? Examining the role of robot size on perceptions of mistreatment. In *2016 25th IEEE international symposium on robot and human interactive communication (RO-MAN)* (pp. 1071–1076). IEEE Press.

Luria, M., Pusateri, J., Oden Choi, J., Aronson, R., Yildirim, N., & Steenson, M. W. (2020a, July). Medieval robots: The role of historical automata in the design of future robots. In *Companion Publication of the 2020 ACM on Designing Interactive Systems Conference* (pp. 191–195).

Luria, M., Sheriff, O., Boo, M., Forlizzi, J., & Zoran, A. (2020b). Destruction, catharsis and emotional release in human-robot interaction. *ACM Transactions on Human-Robot Interaction (THRI)*. https://doi.org/10.1145/3385007

Lynn, J. (2020, January 6). *Why do I want to kick those cute little food-delivery robots? The Bold Italic*. https://thebolditalic.com/why-do-i-want-to-kick-those-cute-little-food-delivery-robots-ba0555a144d9

MacDorman, K. F. (2005, December). Mortality salience and the uncanny valley. In *5th IEEE-RAS International Conference on Humanoid Robots, 2005*. (pp. 399–405). IEEE Press.

Mackenzie, R. (2020). Sexbots: sex slaves, vulnerable others or perfect partners? In *Robotic systems: Concepts, methodologies, tools, and applications* (pp. 1307–1325). IGI Global.

MacKinnon, L. (2019). Coal and steel, goodbye to all that: Symbolic violence and working-class erasure in postindustrial landscapes. *Labor, 16*(1), 107–125.

Maggi, F., Quarta, D., Pogliani, M., Polino, M., Zanchettin, A. M., & Zanero, S. (2017). Rogue robots: Testing the limits of an industrial robot's security. *Trend Micro, Politecnico di Milano, Technical Reports*. http://documents.trendmicro.com/assets/wp/wp-industrial-robot-security.pdf

Manuel, F. E. (1938). The Luddite movement in France. *The Journal of Modern History, 10*(2), 180–211.
Mars, G. (2019). *Work place sabotage*. Routledge.
Mattoni, A., & Teune, S. (2014). Visions of protest: A media-historic perspective on images in social movements. *Sociology Compass, 8*(6), 876–887. https://doi.org/10.1111/soc4.12173
Matyszczyk, C. (2015). Judge rules man had right to shoot down drone over his house. *C/net*. https://www.cnet.com/news/judge-rules-man-had-right-to-shoot-down-drone-over-his-house/
Mayor, A. (2020). *Gods and robots: Myths, machines, and ancient dreams of technology*. Princeton University Press.
Mayoral-Vilches, V., Carbajo, U. A., & Gil-Uriarte, E, (2020). Industrial robot ransomware: Akerbeltz. *2020 Fourth IEEE International Conference on Robotic Computing (IRC)*, pp. 432–435. https://doi.org/10.1109/IRC.2020.00080
McClure, P. K. (2018). "You're fired," says the robot: The rise of automation in the workplace, technophobes, and fears of unemployment. *Social Science Computer Review, 36*(2), 139–156. https://doi.org/10.1177/0894439317698637
McKnight, J. C. (2014, December). Dombots: An ethical and technical challenge to the robotics of intimacy. In *Robophilosophy* (pp. 253–261).
McLinton, S. S., Zadow, A., Neall, A. M., Tuckey, M. R., & Dollard, M. F. (2018). Violence and psychosocial safety climate; quantitative and qualitative evidence in the healthcare industry. In *Violence and abuse in and around organisations* (pp. 39–58). Routledge.
Melson, G. F., Kahn, P. H., Jr., Beck, A., & Friedman, B. (2009). Robotic pets in human lives: Implications for the human–animal bond and for human relationships with personified technologies. *Journal of Social Issues, 65*(3), 545–567.
Meryem, S., & Mazri, T. (2019). Security study and challenges of connected autonomous vehicles. In *Proceedings of the 4th International Conference on Smart City Applications* (pp. 1–4).
Miller, R. J. (1983). The human: Alien in the robotic environment? *The Annals of the American Academy of Political and Social Science, 470*(1), 11–15.
Moore, D., Currano, R., Shanks, M., & Sirkin, D. (2020, March). Defense against the dark cars: Design principles for griefing of autonomous vehicles. In *Proceedings of the 2020 ACM/IEEE International Conference on Human-Robot Interaction* (pp. 201–209). https://doi.org/10.1145/3319502.3374796
Mori, M. (2012, June). The uncanny valley. *IEEE Robots & Automation Magazine, 19*(2), 98–100. http://ieeexplore.ieee.org/xpl/tocresult.jsp?isnumber=6213218&punumber=100

Morris, D. (2015, December 21). As robots muscle in, Rotterdam port workers vote to strike. *Fortune*. https://fortune.com/2015/12/21/rotterdam-port-robots-strike/

Moses, J., & Ford, G. (2021). Is 'Spot' a good dog? Why we're right to worry about unleashing robot quadrupeds. *The Conversation*. https://theconversation.com/is-spot-a-good-dog-why-were-right-to-worry-about-unleashing-robot-quadrupeds-160095

Mrug, S., Loosier, P. S., & Windle, M. (2008). Violence exposure across multiple contexts: Individual and joint effects on adjustment. *American Journal of Orthopsychiatry, 78*(1), 70–84.

Murrer, S. (2020, March 11). Starship robots avoid mishap by 'screeching' when people try to rob them. *Milton Keynes*. https://www.miltonkeynes.co.uk/news/people/starship-robots-avoid-mishap-screeching-when-people-try-rob-them-milton-keynes-2446856

Nevett, J. (2018, January 24). UK's first sex doll brothel faces backlash as prostitutes fear 'going out of business'. *Daily Star* (Online). https://www.dailystar.co.uk/news/latest-news/676702/uk-sex-doll-brothel-lovedoll-prostitutes-backlash-business-gateshead

Nichols, G. (2017, October 2). Sex robot molested and destroyed at electronics show. *ZDnet*. https://www.zdnet.com/article/sex-robot-molested-destroyed-at-electronics-show/

Nomura, T., Kanda, T., Kidokoro, H., Suehiro, Y., & Yamada, S. (2016). Why do children abuse robots? *Interaction Studies, 17*(3), 347–369.

Neff, G., & Nagy, P. (2016). Automation, algorithms, and politics talking to Bots: Symbiotic agency and the case of Tay. *International Journal of Communication, 10*, 4915–4931.

Neuman, J. H., & Baron, R. A. (1998). Workplace violence and workplace aggression: Evidence concerning specific forms, potential causes, and preferred targets. *Journal of Management, 24*(3), 391–419.

Oppenheimer, A. (2018, June 1). Las Vegas hotel workers vs. robots is a sign of looming labor struggles. *Miami Herald*. https://www.miamiherald.com/news/local/news-columns-blogs/andres-oppenheimer/article212375634.html

Oravec, J. A. (2017a). Experiments in protesting: Virtual sit-ins, hacktivism, electronic civil disobedience, and other technology-mediated protest before the advent of social media. In *'Voices of Dissent': Social Movements and Political Protest in Post-war America*. Rothermere American Institute, University of Oxford. https://www.rai.ox.ac.uk/sites/default/.../voices_of_dissent_programme_12.5.17.pdf

Oravec, J. A. (2017b). Kill switches, remote deletion, and intelligent agents: Framing everyday household cybersecurity in the Internet of Things. *Technology in Society, 51*, 189–198.

Oravec, J. A. (2019). Artificial intelligence, automation, and social welfare: Some ethical and historical perspectives on technological overstatement and hyperbole. *Ethics and Social Welfare, 13*(1), 18–32.

Paerregaard, K. (2019). Grasping the fear: How xenophobia intersects with climatephobia and robotphobia and how their co-production creates feelings of abandonment, self-pity and destruction. *Migration Letters, 16*(4), 647–652.

Palmer, A. (2019, January 21). Amazon reveals 'anti-robot vest' to protect workers from giant machines smashing into them. *Daily Mail*. https://www.dailymail.co.uk/sciencetech/article-6617173/Amazon-reveals-anti-robot-vest-protect-workers-giant-machines-smashing-them.html

Pankewitz, C. (2017). Automation, robots, and algorithms will drive the next stage of digital disruption. *Phantom ex Machina* (pp. 185–196). Springer International Publishing.

Pearson, G. (1979). Resistance to the machine. In H. Nowotny & H. Rose (Eds.), *Counter-movements in the sciences* (pp. 185–220). Springer.

Pearson, Y., & Borenstein, J. (2021). The impact of robot companions on the moral development of children. In *Engineering and philosophy* (pp. 237–248). Springer.

PenzeyMoog, E., & Slakoff, D. C. (2021). *As technology evolves, so does domestic violence: Modern-Day tech abuse and possible solutions*. Emerald Publishing Limited.

Perales Gomez, A. L., Fernández Maimó, L., Huertas Celdran, A., Garcia Clemente, F. J., Gil Pérez, M., & Martínez Pérez, G. (2021). SafeMan: A unified framework to manage cybersecurity and safety in manufacturing industry. *Software: Practice and Experience, 51*(3), 607–627.

Peters, R., Broekens, J., Li, K., & Neerincx, M. A. (2019, September). Robots expressing dominance: Effects of behaviours and modulation. In *2019 8th International Conference on Affective Computing and Intelligent Interaction (ACII)* (pp. 1–7). IEEE.

Prianto, Y., Sumantri, V. K., & Sasmita, P. Y. (2020, May). Pros and cons of AI robot as a legal subject. In *Tarumanagara International Conference on the Applications of Social Sciences and Humanities (TICASH 2019)* (pp. 380–387). New York: Atlantis Press.

Purtill, C. (2020, February 25). What do we look for in a good robot colleague. *BBC News*. https://www.bbc.com/worklife/article/20200219-what-do-we-look-for-in-a-good-robot-colleague

Ramirez, J. J. (2020). *Against automation mythologies: Business science fiction and the ruse of the robots*. Routledge.

Ranabhat, B., Clements, J., Gatlin, J., Hsiao, K. T., & Yampolskiy, M. (2019). Optimal sabotage attack on composite material parts. *International Journal of Critical Infrastructure Protection, 26*, 100301.

Randazzo, R. (2018, December 11). A slashed tire, a pointed gun, bullies on the road: Why do Waymo self-driving vans get so much hate? *Arizona Republic.* https://www.azcentral.com/story/money/business/tech/2018/12/11/waymo-self-driving-vehicles-face-harassment-road-rage-phoenix-area/2198220002/

Ras, I., & Gregoriou, C. (2019). The quest to end modern slavery: Metaphors in corporate modern slavery statements. *Anti-trafficking Review*, (13), 100–118.

Rehm, M., & Krogsager, A. (2013, August). Negative affect in human robot interaction—Impoliteness in unexpected encounters with robots. In *2013 IEEE RO-MAN* (pp. 45–50). IEEE.

Reyes, M. E., Meza, I. V., & Pineda, L. A. (2019). Robotics facial expression of anger in collaborative human–robot interaction. *International Journal of Advanced Robotic Systems, 16*(1). https://doi.org/10.1177/1729881418817972

Reynolds, C., & Ishikawa, M. (2006, October). Robot trickery. In *International workshop on ethics of human interaction with robotic, bionic, and AI systems: Concepts and policies.*

Richardson, K. (2015). *An anthropology of robots and AI: Annihilation anxiety and machines* (Vol. 20). Routledge.

Richardson, T., Sur, I., & Amor, H. B. (2020). "Should robots feel pain?"—Towards a computational theory of pain in autonomous systems. In *Robotics research* (pp. 91–99). Springer.

Ringler, J., & Reckter, H. (2012, February). DESU 100: About the temptation to destroy a robot. In *Proceedings of the Sixth International Conference on Tangible, Embedded and Embodied Interaction* (pp. 151–152).

Rodell, R. (2020). We burn our robots in Africa; they keep trying to steal our jobs. *The South African.* https://www.thesouthafrican.com/technology/how-robots-are-affecting-africa/

Rolenc, J. M. (2020). Technological change and innovation as security threats. In *SHS Web of Conferences* (Vol. 74, p. 02015). EDP Sciences.

Romero, S. (2018, December 31). Wielding rocks and knives, Arizonans attack self-driving cars. *New York Times.* https://www.nytimes.com/2018/12/31/us/waymo-self-driving-cars-arizona-attacks.html

Roosevelt, M. (2019, April 16). Port of L.A. automation vote is delayed after dockworkers protest. *LA Times.* https://www.latimes.com/business/la-fi-los-angeles-port-truck-automation-20190416-story.html

Roscigno, V. J., Hodson, R., & Lopez, S. H. (2009). Workplace incivilities: the role of interest conflicts, social closure and organizational chaos. *Work, Employment and Society, 23*(4), 747–773.

Rossmy, B., Völkel, S. T., Naphausen, E., Kimm, P., Wiethoff, A., & Muxel, A. (2020, July). Punishable AI: Examining users' attitude towards robot punishment. In *Proceedings of the 2020 ACM Designing Interactive Systems*

Conference (pp. 179–191). https://dl.acm.org/doi/abs/10.1145/3357236.3395542

Sabur, R. (2018, March 6). *Robo-rage*: Humans reported to be attacking self driving cars in California. *Telegraph* (UK). https://www.telegraph.co.uk/news/2018/03/06/robo-rage-humans-reported-attacking-self-driving-cars-california/

Sahota, N., & Ashley, M. (2019). When robots replace human managers: Introducing the quantifiable workplace. *IEEE Engineering Management Review, 47*(3), 21–23.

Sainato, M. (2020, February 5). "I'm not a robot:" Amazon workers protest unsafe and grueling conditions. *The Guardian.* https://www.theguardian.com/technology/2020/feb/05/amazon-workers-protest-unsafe-grueling-conditions-warehouse

Sale, K. (1996). *Rebels against the future: The Luddites and their war on the Industrial Revolution: Lessons for the computer age.* Basic Books.

Salvini, P., Ciaravella, G., Yu, W., Ferri, G., Manzi, A., Mazzolai, B., ... & Dario, P. (2010, September). How safe are service robots in urban environments? Bullying a robot. In *19th International Symposium in Robot and Human Interactive Communication* (pp. 1–7). IEEE Press.

Sandberg, A. (2015). Death and pain of a digital brain. *New Scientist, 227*(3038), 26–27.

Sanderson, L. M., Collins, J. W., & McGlothlin, J. D. (1986). Robot-related fatality involving a US manufacturing plant employee: Case report and recommendations. *Journal of Occupational Accidents, 8*(1–2), 13–23.

Schrage, M. (2016). Why you shouldn't swear at Siri. *Harvard Business Review.* https://hbr.org/2016/10/why-you-shouldnt-swear-at-siri

Shao, M., Snyder, M., Nejat, G., & Benhabib, B. (2020). User affect elicitation with a socially emotional robot. *Robotics, 9*(2), 44.

Shelley, M. (2012/1818). *Frankenstein.* Broadview Press.

Sherman, A. (2018, December 29). Now is the time to figure out the ethical rights of robots in the workplace. *CNBC.* https://www.cnbc.com/2018/12/27/now-is-the-time-to-figure-out-the-ethical-rights-of-robots-in-the-workplace-.html

Singer, P. W. (2009). *Wired for war: The robotics revolution and conflict in the 21st century.* Penguin.

Singh, Y. (2020). Workplace deviance: A conceptual framework. In *Analyzing workplace deviance in modern organizations* (pp. 1–22). IGI Global.

Smith, A. (2020, April 15). *Over 50 cell towers vandalized in UK due to 5G Coronavirus conspiracy theories.* https://au.pcmag.com/digital-life/66385/over-50-cell-towers-vandalized-in-uk-due-to-5g-coronavirus-conspiracy-theories

Smith, C., Somanath, S., Sharlin, E., & Kitamura, Y. (2020, March). Exploring interactions between rogue autonomous vehicles and people. In *Companion of the 2020 ACM/IEEE International Conference on Human-Robot Interaction* (pp. 453–455).

Smokowski, P. R., & Evans, C. B. (2019). Playground politics, power, and privilege in the workplace: How bullying and harassment impacts employees. *Bullying and victimization across the lifespan* (pp. 187–204). Springer.

Sourdin, T. (2018). Judge v. Robot: Artificial intelligence and judicial decision-making. *University of New South Wales Law Journal, 41*(4), 1114–1133.

Sparrow, R. (2020). Virtue and vice in our relationships with robots: Is there an asymmetry and how might it be explained? *International Journal of Social Robotics*, 1–7. https://doi.org/10.1007/s12369-020-00631-2

Spatola, N. (2020, March). Would you turn off a robot because it confronts you with your own mortality? In *Companion of the 2020 ACM/IEEE International Conference on Human-Robot Interaction* (pp. 61–68). IEEE Press.

Spatola, N., Belletier, C., Normand, A., Chausse, P., Monceau, S., Augustinova, M., ... & Ferrand, L. (2018). Not as bad as it seems: When the presence of a threatening humanoid robot improves human performance. *Science Robotics, 3*(21). https://doi.org/10.1126/scirobotics.aat5843

Spence, P. R., Westerman, D., Edwards, C., & Edwards, A. (2014). Welcoming our robot overlords: Initial expectations about interaction with a robot. *Communication Research Reports, 31*(3), 272–280.

Stephens, A. N., Trawley, S. L., & Ohtsuka, K. (2016). Venting anger in cyberspace: Self-entitlement versus self-preservation in #roadrage tweets. *Transportation Research Part F: Traffic Psychology and Behaviour, 42*, 400–410.

Stern, B. J. (1937). Resistances to the adoption of technological innovations. *Journal of the Patent Office Society, 19*, 930–937.

Strait, M., Contreras, V., & Vela, C. D. (2018). Verbal disinhibition towards robots is associated with general antisociality. *arXiv preprint* arXiv:1808.01076

Sweet, P. L. (2019). The sociology of gaslighting. *American Sociological Review, 84*(5), 851–875.

Tan, X. Z., Vázquez, M., Carter, E. J., Morales, C. G., & Steinfeld, A. (2018, February). Inducing bystander interventions during robot abuse with social mechanisms. In *Proceedings of the 2018 ACM/IEEE International Conference on Human-Robot Interaction* (pp. 169–177).

Tangermann, V. (2021). Remorseless man brags about abusing Tesla's self-driving features. *Futurism*. https://futurism.com/remorseless-man-brags-tesla

Tavory, I., & Fine, G. A. (2020). Disruption and the theory of the interaction order. *Theory and Society, 49*, 365–385. https://doi.org/10.1007/s11186-020-09384-3

Taylor, A. (2022, January, 19). Men are creating AI girlfriends, verbally abusing them, and bragging about it on Reddit. *Fortune.* https://fortune.com/2022/01/19/chatbots-ai-girlfriends-verbal-abuse-reddit/

Teo, Y. (2020). Recognition, collaboration and community: Science fiction representations of robot careers in *Robot & Frank, Big Hero 6* and *Humans. Medical Humanities.* https://mh.bmj.com/content/early/2020/04/21/medhum-2019-011744.abstract

Terbrack, J. (2021, March 30). Starship vs. whips: Robots pose potential risk on streets. *Bowling Green Falcon Media.* https://www.bgfalconmedia.com/news/starship-vs-whips-robots-pose-potential-risk-on-streets/article_f313 31b2-91af-11eb-af2b-d7feb69985a3.html

Thompson, C. (2018, April 6). Elon Musk warns that creation of 'god-like' AI could doom mankind to an eternity of robot dictatorship. *Business Insider.* https://www.businessinsider.com/elon-musk-says-ai-could-lead-to-robot-dictator-2018-4

Tierney, M. (2019). *Dismantlings: Words against machines in the American long seventies.* Cornell University Press.

Torrez, A. (2019, August 20). Hayward P.D. searching for person who kicked, damaged robot security guard. *Ktvu.com.* https://www.ktvu.com/news/hayward-p-d-searching-for-person-who-kicked-damaged-robot-security-guard

Trombola, N. (2021, April 26). Bloomfield residents raise concerns about sharing sidewalk space with delivery robots. *Pittsburgh Post-Gazette.* https://www.post-gazette.com/local/city/2021/04/26/kiwibot-personal-delivery-robots-PDD-pilot-program-bloomfield-pittsburgh-liability-concerns-DOMI/stories/202104220188

Turpin, A., Shier, M. L., Nicholas, D. B., & Graham, J. R. (2019). Workplace violence in human service organizations: A qualitative inquiry of team-level dynamics. *Journal of Health & Human Services Administration, 42*(3), 259–304.

Vanman, E. J., & Kappas, A. (2019). "Danger, Will Robinson!" The challenges of social robots for intergroup relations. *Social and Personality Psychology Compass, 13*(8), 1–13. https://doi.org/10.1111/spc3.12489

Vinanzi, S., Goerick, C., & Cangelosi, A. (2019, August). Mindreading for robots: Predicting intentions via dynamical clustering of human postures. In *2019 Joint IEEE 9th International Conference on Development and Learning and Epigenetic Robotics (ICDL-EpiRob)* (pp. 272–277). IEEE Press.

Walsh, E. J. (1986, March). The role of target vulnerabilities in high-technology protest movements: The nuclear establishment at Three Mile Island. In *Sociological forum* (Vol. 1, No. 2, pp. 199–218). Kluwer Academic Publishers.

Walter, E. (2016, April 8). RIT team gets ready for BattleBots show in LA. *Democrat and Chronicle*. https://www.democratandchronicle.com/story/money/2016/04/08/rit-students-engage-battlebots-robotic-combat/82481746/

Waytz, A., & Norton, M. I. (2014). Botsourcing and outsourcing: Robot, British, Chinese, and German workers are for thinking—not feeling—jobs. *Emotion, 14*(2), 434.

Weininger, E. B. (2002). Pierre Bourdieu on social class and symbolic violence. In E. Wright (Ed.), *Approaches to class analysis* (pp. 116–165). Cambridge University Press.

Westerlund, M. (2020). The ethical dimensions of public opinion on smart robots. *Technology Innovation Management Review, 10*(2), 25–36. https://doi.org/10.22215/timreview/1326

While, A. H., Marvin, S., & Kovacic, M. (2021). Urban robotic experimentation: San Francisco, Tokyo and Dubai. *Urban Studies, 58*(4), 769–786.

Whitby, B. (2008). Sometimes it's hard to be a robot: A call for action on the ethics of abusing artificial agents. *Interacting with Computers, 20*, 326–333.

Willison, R., & Warkentin, M. (2013). Beyond deterrence: An expanded view of employee computer abuse. *MIS Quarterly, 37*(1), 1–20.

Winfield, A. F., Winkle, K., Webb, H., Lyngs, U., Jirotka, M., & Macrae, C. (2020). Robot accident investigation: A case study in responsible robotics. *arXiv preprint* arXiv:2005.07474

Winkelman, Z., Buenaventura, M., Anderson, J. M., Beyene, N. M., Katkar, P., & Baumann, G. C. (2019). *When autonomous vehicles are hacked, who is liable?* (No. RR-2654-RC). RAND Corporation.

Winkle, K., Melsión, G. I., McMillan, D., & Leite, I. (2021, March). Boosting Robot credibility and challenging gender norms in responding to abusive behaviour: A case for feminist robots. In *Companion of the 2021 ACM/IEEE International Conference on Human-Robot Interaction* (pp. 29–37).

Wolfert, P., Deschuyteneer, J., Oetringer, D., Robinson, N., & Belpaeme, T. (2020, March). Security risks of social robots used to persuade and manipulate: A proof of concept study. In *Companion of the 2020 ACM/IEEE International Conference on Human-Robot Interaction* (pp. 523–525).

Xu, L., & Yu, F. (2019). Factors that influence robot acceptance. *Chinese Science Bulletin, 65*(6), 496–510.

Yeşiltaş, M., & Gürlek, M. (2020). Understanding the nature of deviant workplace behaviors. In *Organizational behavior challenges in the tourism industry* (pp. 305–326). IGI Global.

Zeller, F., Smith, D. H., Duong, J. A., & Mager, A. (2020). Social media in human–robot interaction. *International Journal of Social Robotics, 12*(2), 389–402.

CHAPTER 9

The Future of Embodied AI: Containing and Mitigating the Dark and Creepy Sides of Robotics, Autonomous Vehicles, and AI

There are incredibly strong contrasts between many dark side and bright side projections of how robotics, autonomous vehicles, and AI will affect humans in their daily existences. Cultural theorist Joseph Campbell (2011) characterized his interactions with intelligent technologies as follows: "I have bought this wonderful machine—a computer. Now, I am rather an authority on gods, so I identified the machine—it seems to me to be an Old Testament god with a lot of rules and no mercy." In contrast, AI pioneer Marvin Minsky (1994) declared "Will robots inherit the earth? Yes, but they will be our children." These entities are being construed alternately as gods with no mercy to being our own progeny. Varieties of "Jekyll and Hyde" contrasts are a common theme in drama and literature (Rose, 1996); the idea that aspects of shadowy, negative robotic traits could be coupled with positive and functional ones in some way is readily formulated given the decades of science fiction and theatrical treatment. The "good cop/bad cop" negotiating team notion is comparably designed as a dramaturgical device to reflect how different aspects of an entity can "work together but be in opposition" (Brodt & Tuchinsky, 2000). Despite the considerable negative legacies of and immediate outcomes with robotics and AI, many development and implementation initiatives continue to insert these technologies into everyday life with few mitigation efforts for those who face concerns or even fears or trauma. In some cases, these creepy and fearful legacies are

© The Author(s), under exclusive license to Springer Nature Switzerland AG 2022
J. A. Oravec, *Good Robot, Bad Robot*, Social and Cultural Studies of Robots and AI, https://doi.org/10.1007/978-3-031-14013-6_9

instrumental to the robots' impacts (as with various sex, security, and military robots) and are used opportunistically to enhance their effects. Many evangelistic and advocacy efforts to promote robots, autonomous vehicles, and AI have produced confusion rather than clarity when it comes to critically important and even life-and-death trepidations. This book presents the case that individuals be allowed some psychological and physical distance from robots, autonomous vehicles, and AI when feasible—and be empowered to be free from pushed and compulsory applications of these technologies as well as from the "quartering" of these entities in their living spaces (as described in the chapter on military robots).

The glowing projections and dreams of talented scientists and entrepreneurs have stimulated the development of robotics and AI over the past decades (as related in Kang, 2011, and elsewhere in this book). These compelling positive visions and rhetorical flourishes indeed can inform us but also may distract us from understanding and mitigating the darker and creepier sides of the technologies involved. Efforts to idealize, deify, and mystify AI can make efforts to utilize robotics, autonomous vehicles, and related entities to their best human advantage more difficult (as related in Heffernan, 2020 and elsewhere). This concluding chapter provides some optimistic, "good robot" aspects along with the negative, exploring some of the more hopeful projections involving robots, autonomous vehicles, and AI in light of these technologies' more unsettling legacies and immediate impacts, as well as in their future directions (such as in the metaverse). It also analyzes the damages that malicious, neglectful, or uncritical uses of robotics and other AI-enhanced entities can have on individuals and societies; it also examines the dysfunctional impacts of the opportunistic use of human anxieties, fears, manipulations, and deceptions in these technological realms by some commercial and governmental forces. For example, some of the anxieties associated with robotics and AI may be generated by the various well-publicized human–robot or human-AI contests and demonstrations that are established by some corporations. The organizations involved are often effectively engaged in demonstrating that human input is less valuable than robotic input, in showing that robots "outclass" humans in some ways, and in supporting AI hype as a way of expanding specific technological influences and control structures (as related in previous chapters). Other forms of negative impact may be less obvious on the surface: the "dark patterns" approaches that manipulate users into participating in interactions about which they are deceived or the full consequences of which they do not understand have unfortunately migrated to the development of robots and automated vehicles. Opportunistic approaches to robotic and AI initiatives also include the fostering of addictive properties of some consumer robots, such as sex and domestic robots. Many individuals who are currently addicted to smartphone usage may soon find other available objects in sex robots. We may indeed be living with dark and creepy sides of robotics and AI for some time, especially as these negative aspects become normalized and the opportunistic utilization of these dark dimensions is shown to have economic and social utility for some governments and organizations.

This book proposes ways of containing the dark and creepy sides of these technologies; however, these various negative dimensions are evolving and today's proposed remedies may not work in the future. A technological "arms race" is transpiring in which various countermaneuvers can effectively nullify safety, security, and wellbeing-related advances. Other methods of containing the "dark side" are potentially more outlandish: the right to be able to limit or avoid contact with robots or AI-enhanced entities may not be something that can be feasibly implemented in some societal contexts, but should be considered seriously in many settings. The notion that individuals should be accepting of people of other races, religions, and genders and interact with them need not extend to robots and AI-related entities, however humanoid they may appear and however deeply embedded they are in societal structures. Compulsory or "pushed" contact with robots can diminish the quality of life in many circumstances, and at least identifying those zones and spaces in which individuals can be "robot free" and opt out of interaction with robots and AI is a start. In comparable ways, passengers and drivers who wish to avoid autonomous vehicles should be allowed the option of human-controlled means of transportation when feasible. Rhetorical and marketing efforts to convince automobile drivers that they are less capable than autonomous vehicle equipment should be replaced by critical and balanced analyses of our societal transportation systems and specific driving options.

Protecting Mental, and Physical Integrity

Many of the interactions with robots and AI-enhanced entities described in this book have substantial implications for individuals' "mental and physical integrity" (Ruggiu, 2018). The European Court of Human Rights has outlined the protection of mental and physical integrity as major dimensions of human rights defense, although issues with the specific technologies of robotic systems have not yet been directly treated. Basic human rights concerns would place the autonomy of employees first in robotic and automated vehicle design, not the ideological, economic, or esthetic preferences of workplace owners or administrators. Interventions that would disrupt the integrity and wholeness of human beings as independent, autonomous entities would certainly run counter to human rights, even though in some senses a number of them can be construed as voluntary. Uses of robots and autonomous vehicles in community and organizational contexts in the US and Europe have indeed largely been framed as voluntary, although increasingly they have compulsory dimensions. For example, interactions with the police and security robots that roam many communities and organizations are certainly not voluntary, and may even include remote lie and cheating detection as well as facial recognition capabilities (Oravec, 2022a, 2022b; Oravec, 2022a, 2022b forthcoming). As robots become smaller (such as the microscopic-sized "swarm" robots described in Olaronke et al., 2020), the lack of visibility involved would make human interactions with robots even less deliberate and transparent. Grace Murray Hopper coined the expression "bug" for computer errors and mishaps, since a tiny moth was identified as the cause of a particular computer malfunction (Schieber, 1987). Swarm robots indeed present another kind of hazard to human beings.

Why not just alert people to potential hazards with robots and AI—won't that be sufficient to contain any possible harm or menace? It will not be easy to be proactive in warning people about the dark sides of robotics and AI. For example, in one kind of robot application, medical experts are not clear as to whether recommending that patients have sex with a robot would be harmful to their mental and physical autonomy; it may be years before the long-term implications of robot sex on individuals' health and wellbeing emerge (Dewitte & Reisman, 2021). Individuals who are faced with friends, employers, and clients who are drawn toward the positive images of robotics and AI (as crafted by marketers and advocates) may run into many obstacles in warning people about or even discussing these dark side issues in their workplace or community settings. Futurist and robotics pioneer J. F. Coates castigated researchers to wait until the ill effects of technologies were more fully realized, and then observe what kinds of mitigations would emerge organically (Coates, 2000). In contrast, this book has taken on a more proactive and future-oriented approach, aiming toward stimulating users and developers to examine the issues as they emerge, develop humane and functional solutions, and project the implications of other related, evolving concerns.

We are currently in the very early stages of robot and autonomous vehicle implementation, despite the fact that some of the basics of robotics have been around for centuries. As these technologies function in increasingly critical and intimate aspects of our lives, the designers, maintainers, and users of robots and AI entities will be encountering many of the dark side concerns outlined in this book. Investors in robotics and AI as well as developers themselves can take substantial steps in mitigating emerging problems by listening to users and taking their fears and concerns seriously. Having blind faith in technological progress, believing that the various problems will "work themselves out" (as outlined by Coates as well as many other developers and engineers) may not be a sound strategy for the long run. The damages that may occur because of our lack of problem diagnosis and timely intervention can be severe. The "collaborations" that we have with machines (according to former US Secretary of State Henry Kissinger) will be "as consequential" but are "less predictable" than nuclear weapons (quoted in Schizer, 2021), making our continued alertness to new and incipient developments imperative.

PROJECTING UTOPIAN AND DYSTOPIAN SCENARIOS

In a concluding chapter of a book (especially one that highlights dark and somber issues), looking forward to the future can provide a form of relief. Futurist approaches have a long intellectual history (Beckwith, 1984; Hencley, 1974); in past centuries, individuals produced utopian and dystopian accounts both to express current political opinions and to project long-range social agendas (Baczko, 1989). H. G. Wells' vision of the future in *The Shape of Things to Come* (1936) presented technologists as heroes and inspired a number of related films and commentaries, as did Edward Bellamy's revelations in *Looking Backward* (1888). In contrast, George Orwell's *1984* (1948) and Aldous Huxley's *Brave New World* (1932) conveyed imageries of a future gone awry and of technologists being part of the problem rather than the solution. Some advances in robotics and virtual reality are closely tied to science fiction literature, such as the writings of Isaac Asimov (1986) and William Gibson (1986). Years ago, former Microsoft CEO Bill Gates projected that "… a robot in every home" would aid in empowering the occupants (Gates, 2007, p. 58), potentially providing solutions to everyday domestic and personal problems. In the past decade, UK futurist Ian Pearson predicted "By 2050 androids will do most of the cooking and will even have rights, they will have to be treated with respect and possibly paid" (Alexander, 2019).

Many of the people reading this book will be active consumers and citizens at the end of this century, so they will eventually find out whether the scenarios below were accurate. Consider the following projections as to how robotics and autonomous vehicles will be part of your future lives:

> By the end of the 21st century, robots will be ubiquitous throughout human societies, particularly in developed countries…. They will replace both white- and blue-collar workers, provide child- and elder-care, and serve as personal assistants in the home. Humanoid robots will provide more than information and labor. Computer engineers are already working on robots that not only physically look human but also resemble humans in terms of emotional and sexual capacities. If human beings embrace humanoid robots as their intimate companions, perhaps even wed them as marriage partners, the consequences would include the decline in the birth rate, particularly in developed countries, which contribute more carbon emissions per capita to global warming than developing nations. (McBride, 2021, para. 3)

Yet another futuristic scenario outlines the potential for how self-driving vehicles will affect everyday life:

> The year is 2096. Self-driving cars and trucks have reshaped commutes, commerce, and the inner-workings of cities. Artificially intelligent systems have placed sophisticated computer minds in sleek robot bodies. Cognitive assistants—running on an intricate network of sensors monitoring humanity's every move—help finish people's sentences, track and share their whereabouts in real-time, automatically order groceries and birthday gifts based on complex personalized algorithms, and tell humans where they left their sunglasses. Robots have replaced people in the workforce

en masse, claiming entire industries for machine work. There is no distinction between online and offline. Almost every object is connected to the Internet. (LaFrance, 2016, para. 1)

The previous scenarios emphasize the "replacement" of human input and effort, a common theme in projections of the future. However, a variant on these utopian scenarios involving robots and other AI-enhanced entities would have them used in ways that enhance, energize, and augment human activities and processes, not replace humans. Their operations would be transparent and explainable and users would be empowered to make needed changes to fit their situations and environments. The entities would work without manifesting demographic or cultural biases; also, whatever changes in environments they would need to operate could be made in keeping with human needs for the utilization of the spaces involved. However, some of these "replacement" ideals and perspectives are unfortunately reflected in many marketing advertisements and even research grant proposals, in which interaction with robots and AI is often characterized as delightful, unproblematic, and completely autonomous, without some strategy as to how to achieve the ideals.

Ideal scenarios can be enlightening to project (especially with new and emerging technologies), but they are often just fanciful undertakings. Many decades of negative images, unsettling narratives, and real-life mortalities involving robots and AI are not easily dismissed or erased. They often play roles in how these technologies are characterized in everyday utilization as well as how research and development efforts are conducted. Analysis of expert opinion by the Pew Research Center shows that there is substantial concern that "ethical AI design" will not be "broadly adopted as the norm within the next decade" (Rainie et al., 2021), which warns us to be vigilant for dark patterns and other opportunistic efforts to exploit or even develop new and unsettling dimensions of robotics and AI. Positive projections of the future often assume that various solutions for problems will be ready or close at hand—for example, that cybersecurity solutions will be available and feasibly implemented, and have minimal "technological debt." Technological debt occurs when developers project that certain security measures will emerge for their initiatives in time for implementation, when those measures are not yet feasible or even conceivable (Oravec, 2017). Some "virtue-centered" approaches focus on the positive aspects of robotic development, while leaving many of the critical dimensions for further exploration. For example, Donhauser (2019) presents environmental robot virtues in this framework, presenting a discussion that "has been almost entirely positive, in that I have discussed the potential benefits of different sorts of environmental robots without casting much light on the potential risks and dangers they could present" (p. 191). A more critical perspective might indeed be less enticing for businesses and grant funders, but could underscore needed changes.

Dystopian scenarios are also common with robotics and AI: for instance, computer scientist Bill Joy's (2020) statement resonates in many critiques of these technologies: "Our most powerful 21st-century technologies- robotics, genetic engineering, and nanotech- are threatening to make humans an endangered species." Some computer scientists and public policy leaders have countered this pessimistic approach (Bringsjord, 2008), often with arguments that society has met comparable technological challenges in the past. However, it is doubtful that the perceptions and realities of the dark and creepy sides of robotics and AI will be eradicated or even significantly diminished soon. Some of the aspects of these dark sides may indeed be "contained"—identified and understood to some extent and their negative effects mitigated. We may learn, for example, how to better identify rogue robots and educate people in the workplace to be wary of them. The malleability and unpredictability of robotic and other AI-related entities, as well as the speed at which new AI-enhanced technologies are emerging, makes any gain only a temporary one as new technological iterations and related issues emerge. As discussed in this book, robots and other autonomous entities can serve to elicit negative responses and dysfunctions on the part of those who work and play with them, and even more disturbing personal and societal issues may emerge as more complex and opaque applications are developed and implemented.

The notion of robots as magical or enchanted entities with the capability to capture humans' "desirous senses of wonderment and anticipation" (Belk et al., 2021, p. 25) may introduce some positive feelings into everyday life but can also backfire. Associations of robotics and AI with religious themes and imageries can be even more troubling, with spiritual associations of the Singularity as well as demonic associations presenting problematic examples. Heffernan (2020) wrote of the "dangers of mystifying artificial intelligence and robotics," which are only expanding as robots and automated vehicles are put to expanded uses in workplaces and communities. Misplaced magical thinking and enhanced empathic feeling toward robots and autonomous entities can dislodge us from our rational capabilities as well as our gut senses of how to handle potentially rogue and manipulated ("dark pattern") entities. Societies are generally very capable of creating magical and spiritual realms, and of keeping individuals in unvigilant and infantile states in relation to those realms, especially when individuals are otherwise overworked or stressed (Körtner, 2016). The following poem has served for decades as a guide as to how many individuals may eventually consider themselves not as autonomous beings but as seemingly fortunate (through infantilized) entities watched over by robotic "machines of loving grace":

> I like to think
> (it has to be!)
> of a cybernetic ecology
> where we are free of our labors
> and joined back to nature,
> returned to our mammal
> brothers and sisters,
> and all watched over
> by machines of loving grace. (Brautigan, 1968)

As the robot-inflicted deaths described in the previous chapters illustrate, such states of childish bliss can serve to extricate many individuals from some of their "natural" reactions and instincts and could be exceedingly dangerous in relation to robotics and AI. This book has provided a warning to those individuals as well as the robot developers who have moved them into these states. Designers and developers generally want optimal outcomes for the technologies that they produce; those who create robots and other AI applications usually want their conceptions to play useful roles in society, not become anti-social and malicious entities that crush skulls and ruin lives (Oravec, 2021). Through public policy modifications as well as moral persuasion, those who create robots and other AI-enhanced entities need to be reminded of the potential "dark side" consequences of their efforts.

Issues with the Opportunistic Use of the Dark Side

This book has explored only a few of the ways in which the dark sides of technologies have been opportunistically utilized. Through the past decades, many organizations have exploited the fears of individuals about their personal appearances, social standings, and personal safety; they have marketed and disseminated cosmetics, drugs, automobiles, and security systems in opportunistic ways even when they were aware of the potential hazards involved (Laham, 2020). Robots and AI provide a new set of fears and dangers to exploit. Some developers and disseminators of robots and AI-enhanced entities do more than extoll the positive advantages of their products in solving problems for individuals and organizations. For example, robot- and AI-related narratives that can generate fear and anxiety on the part of individuals are readily available, and can be used by developers or organizations to motivate some sort of human behavior or forestall other behaviors. The amorphous and diffuse kinds of fears that are associated with the police and military use of robotics may have a "chilling effect" on everyday activities, since individuals will not know whether or how the robots or AI-enhanced entities may intervene in any particular situation. As mentioned in previous sections the sense of intellectual and spiritual inferiority in relation to robotics and AI can itself be disempowering and perhaps immobilize the critical faculties of individuals who need to make precise decisions as to how far to trust these AI-enhanced entities. The situation with the trust of autonomous vehicles is comparable, with a number of popular benchmarks for determining the fit between human drivers and self-driving vehicles providing potentially-disempowering contexts (Nees, 2019).

Constraining the opportunistic uses of fear and anxiety in the realm of robotics and AI does not mean that people should not be warned about potential hazards in strong and specific manners. Narratives and images that describe disasters and injuries with robots and autonomous vehicles may be an appropriate way of alerting people to potential hazards (although they are also likely to engender fear). Your self-driving vehicle may indeed not perform as expected and the robot that is cleaning your floors or monitoring your children could be sending your personal information to marketers or even the police. The dark and creepy sides of self-driving automobiles may be attractive to some individuals, though, as they attempt to intimidate others on the roadway; the prospects of creepy "headless drivers" have already been explored in popular and academic folklore literature through the past decades (Rodger, 1965). As described in this book, potential security breaches can make nearly any robot or autonomous vehicle untrustworthy, changing its character from a helpful agent to an attacking menace. Characterizing "robots" as entities without considering their malleability can overlook some of their permeability to malicious forces as well as their inherent unreliability. Dark side narratives and images can also be utilized to generate a false sense of personal empowerment, as in the sex robots that can be personalized to take on the characteristics of certain celebrities, acquaintances, and even despised individuals. These senses of empowerment are linked with the magical aspects of robots and AI-enhanced entities, in which interactions with these entities can be perceived to be amplified in some mystical way and transformed in significance.

Opportunistic use of robot- and AI-related anxieties might put organizations in a stronger position in relation to employees. Specific human biases can be fostered through the use of these technologies, despite (or perhaps as a result of) the efforts of developers. For example, with robotic entities that have a metaverse connection, Duan et al. (2021) declare that the "metaverse should be a realistic society with more direct and physical interactions," and predict that "while the concepts of race, gender, and even physical disability would be weakened, which would be highly beneficial for society" (p. 153). However, a "realistic society" has some aspects of bias and discrimination, and metaverse developers have yet to explain how current inequities would be eliminated in this new terrain. The metaverse infrastructure presents a corporate-controlled platform that imposes an opaque layer between individuals and their robotic or AI tools, making the control structures and information collection involved less transparent. Organizations can also utilize various other "dark side" aspects of robotics to their advantage, emphasizing the fear of robotic replacement to their employees even when it is not a practical option. Pew Research Center surveys across a number of nations show that "Publics are generally more worried than hopeful about the prospect that robots and computers may one day be able to do much of the work done by humans today. Their greatest concern is that automation will make it harder for ordinary people to find jobs" (Wike & Stokes, 2018). The rhetoric of "creative disruption" and the "coming robot revolution" can serve to create anxieties for workers about their jobs and their futures, which can affect their abilities to plan for their futures and fight for better working conditions.

Robotic Advocacy: Promoting and Defending Robots and AI

The scope of robotic and AI advocacy is broadening as potential applications emerge, as is the spectrum of feelings about robotic entities among researchers and developers. Critical awareness of these feelings and reflectiveness about how they can affect R&D efforts is needed in order to ensure that the initiatives are implemented and analyzed with some objectivity. The job title of "technology evangelist" has taken on widespread recognition and even popularity for decades (Beatty & Gordon, 1991). Organizations have promoted the "missionary work" of their members and customers, often using such terminology as "conversion, proselytization, devotion, belief, and other religious terms" to support certain organizational processes or technologies (Massa et al., 2017, p. 461). However, efforts toward supporting and defending robots or other technologies (e.g., in a "team robot" configuration) can serve to distort the development and application of the technologies by diminishing the discussion of discordant observations and dissenting views. Techno-utopians are extending the promotion of a particular set of technologies to the ideology that technological progress is inevitable and will ultimately result in utopia. Dinello (2021) has described how many "techno-utopians fervently believe that technological progress will lead to perfection and immortality for the posthuman, cyborg descendants of a flawed, inevitably extinct humanity" (p. 1). These kinds of radically-optimistic beliefs and their rhetorical expressions have considerable real-world economic and social ramifications. For example, "passion from die-hard evangelists" that is often associated with AI has often pushed AI to the "peak of the hype curve" (Santosh & Gaur, 2021, p. 65).

The rhetoric of technological evangelism is often laced with science fiction as well as futurist projections from pioneering scholars and researchers; it is often immoderate, even giddy in tone and extravagant in its enthusiasm about the future. It is sometimes negative about human characteristics and foibles in the attempt to defend and support specific technologies. The Susan Calvin that Isaac Asimov wrote about indeed preferred robots over humans and made a number of direct statements to that effect (Asimov, 1940). Many efforts to endorse and defend robotics and AI are indeed well within the bounds of acceptable levels of enthusiasm and creative energy. However, many are driven by excessively opportunistic, careerist, and promotional incentives that are all too easily injected into research and development initiatives. This book has oversampled from the more critical robot and AI literature, producing a number of research efforts and reflections that present serious and negative viewpoints about technological futures; however, many more are needed. For example, Lacković (2021) describes the kinds of rhetorical support that some algorithms are being given by their developers and implementers as a form of "celebrity," which serves to defend algorithm utilization in specific real-world contexts without examining relevant assumptions and underpinnings.

The extensive and growing levels of advocacy that seek to reduce discourse on robotic and AI hazards because they may somehow impede the acceleration of robotic and AI developments can be dangerous. Scientists and implementers have increasingly played public roles in being apostles for their creations, including nuclear power advocates, whose strong supports of their technologies has apparently often diminished their credibility as experts in public discourse (Ramana, 2018). Scharrer et al. (2017) have shown that science and technology can be presented in non-nuanced and general ways, eroding public trust. The engineering of pro-robot or pro-autonomous vehicle human behavior might be seen as needed in corporate and governmental circles because of the perceived slowness in their rate of adoption; the perception that robotic and AI implementations have been slow to take root can affect economic investment in these technologies and also rush technologies to the market without significant regard for safety and security features. In a comparable manner, the personal computer of the 1980s was construed as an "ally" of the user through marketing and framing of the technologies, which backfired somewhat when viruses and other forms of security breaches emerged as factors in workplaces and homes, often causing large-scale damage (Harris, 1994). The "friendly home PC" was transformed through these security invasions as something more sinister, for which high levels of human vigilance was required.

Some of the advocates and evangelists for robots and AI have focused on fostering empathy and other positive emotions toward these entities, although others have adopted other strategies for promoting them, including negative sanctions for non-compliance. Moral concern for robots as human-produced artifacts is growing on a variety of levels (Nomura et al., 2019); for example, many robots could be given the kinds of ethical status accorded to artworks, given their deep meanings to some members of their societies (Baker, 2020). The notion that a robot would be granted a comparable set of rights as humans has met with some opposition:

> Once we see robots as mediators of human being, we can understand how the 'robots rights' debate is focused on first world problems, at the expense of urgent ethical concerns, such as machine bias, machine elicited human labour exploitation, and erosion of privacy all impacting society's least privileged individuals. (Birhane, A., & van Dijk, 2020, p. 207)

Building human empathy for robots and autonomous vehicles is one mode of robotic and AI advocacy. Some of the advocates for robotics and AI have projected that the era in which robots will have feelings is approaching; Ray Kurzweil predicted in his 2005 book *The Singularity Is Near* that a robot would attain sentience by about 2045. Some robot advocacy approaches are rooted in the idea that robots and other AI-enhanced entities should be treated as forms of personage, and given treatment from this perspective, although Bisconti Lucidi and Nardi (2018), among others, have framed this "robot as personage" approach as an "hallucinatory danger." In some of these approaches, humans who disparage robots would be construed in a manner similar to the ways that individuals often are who exhibit disparagement toward humans of certain racial or gender affiliations. Developers, implementers, and users of AI-enhanced entities should be aware of their assumptions about these entities' place in the social order so that they can obtain some clarity on issues of basic robotic (and human) treatment.

Establishing Open Communication About Robots and AI

For many years, people were aware of the extent of automobile deaths but did little. Consumer advocate Ralph Nader (1965) wrote *Unsafe at Any Speed*, breaking the silence and starting international discourse about how cars could be designed to operate more safely. Decades later, we have indeed made dramatic changes in the safety levels of automobiles.

However, most societies still accept large numbers of fatalities as part of the construction of supposedly-advanced transportation systems, just as they accept the numbers of deaths-by-robot in industrial settings as well as deaths by autonomous vehicle. The kinds of compromises and normalizations that have been made with many other technologies are still being wrought for robots and AI. The numbers and extents of these compromises may lessen our abilities to face important issues that deal with the survival of human beings and the maintenance of our societies. Liu (2018) contends "it may be that the dazzle of AI (as with other emerging technologies) in the context of existential risks is crippling our ability to respond" (p. 25). Other critics construe these technologies as intruding in some essential ways on our standings as human beings, with "machines as unremitting intrusions on human and social life. …with slowly evolving legislation and court decisions that remediate inroads on our humanity" (Zeller et al., 2019, p. 15), calling for an end to technological compromises.

Much of the communication about the negative aspects of robotics and AI has taken place in the form of anti-robot attacks as well as job resignations from places that do R&D on these matters. Assaults on robots of various sorts are continuing, many of which have endangered the lives of other humans as well as created specific physical damage to the robots and AI-enhanced entities themselves. Acts of technological resistance can include large-scale art projects and protest efforts, or the simple (and generally inappropriate) act of just kicking a robot. Acts of resistance can also be small as not using a robot in a particular work situation or in using it incorrectly (Mitchell, 2018; Salzmann-Erikson & Eriksson, 2016). The fact that some engineers and researchers are leaving their jobs over aspects of the technologies they are developing has inspired societal discourse as well as created specific organizational changes, even in the often-opaque realm of military technologies and arms control (Bolton & Mitchell, 2020). For example, newspaper articles with titles such as "Ex-Google worker fears 'killer robots' could cause mass atrocities" (McDonald, 2019) have generated discussion over otherwise hidden or obscure corporate ventures; the individual who quit his job over the military drone project he worked on warned that "AI might also accidentally start a war" (para. 1). As many individuals begin to analyze their own feelings and viewpoints toward robotics, autonomous vehicles, and AI, new approaches for expressing their perspectives may emerge, as well as new modes of collective activity. Several decades ago, such professional groups as the Computer Professionals for Social Responsibility (CPSR) and the Electronic Frontier Foundation (EFF) emerged (as described in a previous chapter), and in future collaborative activity of this kind may take on new forms as the intimate and pervasive impacts of robotics and autonomous vehicles expand.

Open communication about negative robot and AI potentials by users and implementers may potentially help these issues, if it can be conducted without exacerbating fears of being labeled as "anti-tech" or as "Luddites." Rather than being ignorant about technology, many users have taken strong roles in technological development efforts, and the resulting collaborations can result in attitudinal and ethical changes as well as improved products. For example, involvement by users in the prototyping process can serve to alter attitudes toward AI-enhanced entities (Reich-Stiebert et al., 2019). Closer interaction between developers and users might forestall some of the issues of the kind that personal computers ran into in the 1980s and 90 s, in which viruses produced critical damages for a poorly-informed and frightened public. New and unforeseen dark side concerns will indeed emerge in the years to come, and open communication in various managerial and public policy forums could forestall some ill effects rather than waiting for individuals to act out in frustration and fear. With personal computers in past decades, responsibility for antivirus and other security protection was largely placed on the shoulders of individuals (often computer novices). With robotics and autonomous vehicles, such security efforts need to be moved toward a wider, better resourced, and more informed level, with corporations and government agencies taking stronger positions. Expert testimony and other activity can play important roles in these efforts by attempting to focus public attention on the matters involved as well as clarifying the relevant issues (Hancock, 2019).

Development of Proactive Initiatives and Ethical Principles Concerning Robotics and AI

As outlined in this book, designers and developers of robotics and AI-enhanced entities are somewhat limited in what they can do to affect how these entities are utilized in real-world contexts; they can often be modified by consumers at various stages in their use. A number of initiatives are emerging that explore how to deal with robots and autonomous vehicles in ways that mitigate the negative social and economic outcomes analyzed in this book (Abnet, 2020; Black, 2019; Wang & Siau, 2019; Whitby, 2008). Some proactive approaches include workers (Smids et al., 2020) as well as developers and implementers. Some involve the education-related robots used in schools (Newton & Newton, 2019). These initiatives are encountering numerous advertisements, marketing campaigns, and training efforts that are designed to steer people away from critical thinking concerning AI-enhanced entities. Many of these commercial efforts frame robots and autonomous vehicles in positive and friendly perspectives so as to improve their levels of product and service consumption (van Pinxteren et al., 2019). Such persuasive efforts may influence individuals in overlooking the potential ills concerning these entities, ills that can debilitate them in their efforts to use the entities effectively and with consideration of human wellbeing. In relation to the use of military and police robots, civilian oversight of some sort can provide needed insights and awareness of emerging technological developments: such a program "will be key to maintaining a unbiased analysis of aggregated data" concerning killer robots and related initiatives (Jackson, 2020, p. 521). Rights declarations are often used in the resolution of difficult practical and moral conflicts; the formulation of the "right to be secure" in regard to the intrusion of robots and other AI-enhanced entities is an essential dimension of the human right to be free from fear of corporate and governmental invasion (Milligan, 2013; Oravec, 2017).

The advent of robotics, AI, and autonomous vehicles has stimulated the development of ethics-themed initiatives by professional groups and from corporations that are involved with related technological efforts. For example, the Centre for the Study of Existential Risk describes itself as "an interdisciplinary research centre within the University of Cambridge dedicated to the study and mitigation of existential risks" (Schuster & Woods, 2021). Stephen Hawking reportedly joined the Centre in 2013 in part to explore AI-related issues. Some sets of ethical statements have recently emerged from these groups, as in the *National Society of Professional Engineers' statements concerning the ethics of autonomous vehicle development and implementation (revised May 2020):*

> Because the development of autonomous vehicles will have profound impacts on the public health, safety and welfare, it is the position of NSPE that the testing and deployment of such technologies must include a licensed Professional Engineer.
> DISCUSSION:

The rapid advancement of autonomous vehicle technologies, with the eventual goal to develop driverless or entirely autonomous vehicles presents an exciting phase of the technological advancements of humans. This rapid advancement, however, presents the risks associated with development of an advanced technology to be deployed within an aging infrastructure and also on roads occupied by vehicles without this advanced technology.

The automotive industry estimates that by 2025, as many as 230,000 driverless or entirely autonomous vehicles could be sold each year around the world, potentially swelling to 11.8 million a decade later.

RECOMMENDATIONS:

1. To ensure the health, safety, and welfare of the public is held paramount in this development, testing, and deployment, licensed Professional Engineers must be involved in each step of this process to assure transparency. Licensed Professional Engineers should be employed by autonomous vehicle and autonomous technology developers to provide in-house oversight of the development, testing, and deployment, or consulting a third-party licensed Professional Engineer should be used to verify the safety and security of the technology in development, testing, and deployment. (NSPE, 2022)

The Australian Human Rights Commission's Human Rights and Technology Discussion Paper acknowledges an important role for ethical frameworks. The Commission contends, however, that practical impacts of ethical frameworks have been limited in comparison with legislative initiatives (Santow, 2019).

Few ethical principles characterize an individual's rights not to use certain robots, whether because of the individual's personal fears or because of perceived dangers. The workers' or community participants' rights to refuse to deal with a robot or other AI entity could help to empower them to be critical of robot operations and wary of AI processes rather than falling into a "learned helplessness" state (Seligman, 1972) in which they perceive that they have no recourse. Since robots can dramatically change in character because of manipulation and corruption, such protections are especially needed. This right could also diminish some of the motivation for abusing or destroying robots and AI-related entities, since the individuals involved can avoid dealing with the entity in question. Concerns about autonomous vehicles have comparable themes: "the risks that unscrupulous actors might compromise a robot's ethics are so great as to raise serious doubts over the wisdom of embedding ethical decision making in real-world safety–critical robots, such as driverless cars" (Vanderelst & Winfield, 2018, p. 317).

Sets of ethical principles for robotics development have been wrought as organizations are attempting to express their ethical concerns. For example, the Israeli corporation NICE is a major international provider of cloud and on-premises enterprise software solutions. The company developed the following principles "that are intended to ensure good ethical standards, underlying the robot-human relationship in the workplace":

- Robots must be designed for a positive impact: Robots must be built to contribute to the growth and wellbeing of the human workforce. With consideration to societal, economic, and environmental impacts, every project that involves robots should have at least one positive rationale clearly defined.
- Bias-free robotics: Personal attributes such as color, religion, sex, gender, age, and other protected status are eliminated when creating robots so their behavior is employee agnostic. Training algorithms are evaluated and tested periodically to ensure they are bias-free.
- Robots must safeguard individuals: Careful consideration is given to decide whether and how to delegate decisions to robots. The algorithms, processes, and decisions embedded within robots must be transparent, with the ability to explain conclusions with unambiguous rationale. Accordingly, humans must be able to audit a robot's processes and decisions and have the ability to intervene and redress the system to prevent potential offenses.
- Robots must be driven by trusted data sources: Robots must be designed to act based upon verified data from trusted sources. Data sources used for training algorithms should be maintained with the ability to reference the original source.
- Robots must be designed with holistic governance and control: Humans must have complete information about a system's capabilities and limitations. Robotics platforms must be designed to protect against abuse of power and illegal access by limiting, proactively monitoring, and authenticating any access to the platform and every type of edit action in the system. ("NICE Sets," 2021)

NICE's Head of Robotics Process Automation Oded Karev states that NICE adheres to these principles itself and promotes them with its customers, but they are merely voluntary; customers can simply disregard the rules. He says:

> I have no control over how my customers use the robots. But the first and foremost thing is to make sure that my software supports the ability to execute or follow these rules. Our robots are designed with governance and control in mind, which is key to having an audit trail to see who did what, to have roll-back and version control. Every decision can be drawn back to the human who wrote the rules. (Middleton, 2021)

The frustration of individuals who endeavor to shape the directions of robots, autonomous vehicles, and other AI-enhanced technologies toward human ends is often apparent. Major technological shifts are likely as new developers and methodologies emerge, creating a moving target for those concerned about robots and AI's ethical and social implications. For individuals who wish to counter these entities through legal or public policy-related pressures, their changing technological forms as well as the ownership statuses of the systems involved can put frustrating barriers in the way of modification and reform. Social shifts are likely as well: the expansion of "arms races" involving robotics is already occurring as individuals find and share ("crowdsource") various ways to manipulate and alter robots and automated vehicles or subvert lie detection systems. Technology-supported resistance can be a factor in subsequent design iterations; efforts to counter and protest perceived technological intrusions have had sustained impact on the overall direction of technological development, with various surveillance mechanisms serving as challenges to talented and persistent individuals. The technological initiatives described in this book are being evaluated from an assortment of perspectives and standards: for example, Majeed, Baadel, and Ul Haq (2017) contend that these initiatives can either be considered as "global triumph" or as "exploitation" as societal norms adjust to rapid changes in information and communications technology.

A single book of this size cannot cover in depth all of the emerging dark side possibilities linked with robotics, autonomous vehicles, and AI; it certainly cannot address all of the international variations and iterations with their considerable legal and economic impacts. For example, this book does not delve into human "laziness" potentials to any great extent. Science fiction stories about humans becoming lazy and losing their essential will to achieve in the advent of AI can be found in various books, short stories, and films, and will be explored in my own future writings. I have written a great deal about "cyberslacking," and how technologies can change our relationships with labor and the workplace—and indeed robots, automated vehicles, and other AI-enhanced entities could play roles here (Oravec, 2002; 2018). In yet another theme, cultural variations with robots and AI have already affected robot-human interactions as well as perceptions of AI. In the near future, some national differences may be exacerbated because of dissimilarities on perspectives on robotics and AI, such as variations between the US, Japan, and Thailand (Komatsu et al., 2021; Vichitkraivin & Naenna, 2021).

Yet another focus for future examination is how ubiquitous robots, autonomous vehicles, and AI proliferation are affecting children's development. As described in this book, adults have already been exposed to a wide assortment of positive and negative legacies associated with these emerging technologies; children with less exposure may indeed face a different assortment of concerns. As the nuclear family expands to include robotic figures, there could certainly be widespread cultural and personal shifts (as described in Cave & Dihal, 2021). The Ray Bradbury science fiction television show and story "I Sing the Body Electric" (1969) as well as Philip K. Dick's (1955) short story "Nanny" captured many of the issues that are emerging as robots and AI enter the intimate domestic spaces of the home. Science fiction will continue to provide insights as to what the future/ will bring in terms of the impacts of these technologies.

BEYOND THE DARK AND CREEPY: SOME CONCLUSIONS AND REFLECTIONS

Robotic and AI initiatives are branching out into many aspects of societal functioning, and are manifesting a wide assortment of kinds of research and development objectives. Planning for a world filled with autonomous robots and self-driving vehicles has already begun by governments and corporations. Research and development efforts to produce and disseminate robots and AI-enhanced entities are often funded by institutions that will benefit considerably from the sales and applications of these entities and which often do not take significant steps to consider the social and ethical impacts of their ubiquitous proliferation. Some researchers and organizations are making it their goal to produce robots and autonomous vehicles for which humans have considerable emotion (including fondness, empathy, and perhaps even love), which can lead to dangers since a diminution of human vigilance can often result in injury and even death. Others are working hard to perfect police and military robots for which humans have high levels of fear and which can wield substantial control of individuals' behavior.

Eliminating the use of carefully-choreographed "contests" and demonstrations in which humans are pitted against robots in some way can also enable a higher quality of decision making about whether or when robots and AI should be a part of human initiatives. Authorizing workers and community participants to be able to opt out of dealing with robots and AI in various situations can empower them to make critical judgments about their levels of comfort and in doing so inform the efforts of developers and implementers. Individuals who face concerns about robots, autonomous vehicles, and other AI-enhanced entities should be able to contact human beings in the corporations that sell those items and get assistance, not just be forced to handle complex (and often life threatening) technical challenges on their own or with their peers. The kinds of approaches that platform-supporting organizations such as Meta and Alphabet have used to put users at a distance in terms of assistance can be fatal or at least injurious when it comes to robots and autonomous vehicles.

The dimensions of user agency in the realm of robotics and AI are also growing, with the promise of user empowerment; however, such user agency often has unsettling implications as the resulting technological initiatives are used for such activities as simulated child sex or autonomous drone attacks by militia or rogue individuals. Co-development is one form of user agency (as users make substantial changes in the entities), as is resistance (as individuals work to critically oppose the implementations of the technologies in question). Perhaps the near-term resistance to dysfunctional and inappropriate robot and autonomous entity proliferation and implementation will be somewhat constrained and muted, but having some sort of well-considered opposition to the imposition of technologies is vital, be it only a small scattering of academic, literary, or worker-generated voices. Futuristic thinking, trend projections, and scenario development can be of help in engendering insights (Oravec, 1998). Even sarcastic comments, comedies, and parodies can do a great deal to create and maintain a healthy psychological distance between oneself and the potential hazards in one's workplace or community as well as increase one's critical perspective. However, taking robots seriously (especially sex and military robots) is essential to making sure dangerous developments such as child sex robots, rogue autonomous vehicles, and killer robots do not become normalized.

This concluding chapter emphasizes the importance of the critical analysis of the opportunistic use of human anxieties, fears, and delusions relating to robots, autonomous vehicles, and AI. As advanced technologies function in increasingly vital and intimate aspects of our lives, the designers and implementers of robots and AI entities will be encountering many of the "dark side" concerns outlined in this book, with some potentially leveraging them to their advantage. For example, reinforcement of the notion that robots and AI are foreboding and frightening can be used resourcefully if it is applied to make humans wary of robots in dangerous environments or to monitor for hacked or rogue robots. However, high levels of stress and overloads of requirements for monitoring advanced technologies and systems can make individuals prone to mistakes or easy prey for malicious attempts at "social engineering" by hackers. As described in previous chapters, some efforts to contain the dark side of robotics and AI are underway by associating robots and AI further with animate, living creatures (either human or other forms of animal). Anthropomorphism and animistic design may make some robots appear friendlier but may not resolve deeper issues of human vigilance and safety in the realm of human–robot interaction.

In our applications of robotics and other AI-enhanced technologies in workplace and community settings, we as humans are often developing and adopting our own companions and coworkers. We may also be designing our own replacements, our own interrogators, our own jail wardens, and our own executioners. This book is ending with a plea for taking negatively-themed issues seriously when it comes to robotics and AI, despite the fact that many of these concerns are "creepy" to conceptualize and potentially damaging to the economy, as well as unpleasant to talk about. It is often more pleasant (and lucrative in the short term) to project positive, functional scenarios for new technologies than to deal with challenging combinations of negative and dystopic factors. In many organizational contexts in developed nations, expressing a positive interest in technology is often considered a de facto job requirement (in part to avoid the label of "Luddite"), which may lead to workplaces and communities overlooking critically-important factors.

The kinds and numbers of exemplars of how to be open about technological anxieties and fears are expanding. Pioneering organizations such as the Future of Humanity Institute in Oxford University have increased attention to the growing concerns about robots and AI. Some technological and scientific luminaries such as Elon Musk and the late Stephen Hawking have declared their concerns, with Hawking going as far as stating that artificial intelligence "could end mankind" (Cellan-Jones, 2014). Some organizations speaking in formal terms have indeed provided stern warnings about the future, though in many more organizations the economic benefits of robotics and AI have supported congruity with these technologies rather than critical distance. Although the mottos and ethical statements of many high-tech organizations call for enhanced corporate social responsibility, determining which direction to go can be difficult. The "don't be evil" kind of motto that was associated with Google in past decades can provide little support for those who need to make specific decisions about development and implementation in complex systems (Oravec, 2014). Some of the espoused values of the "metaverse" (as outlined by Mark Zuckerberg of Meta Corporation) reportedly include "move quickly," which is akin to the "move fast and break things" slogan of the previous decade (Hamilton, 2022). However, the detailed and collaborative ethical discourse that is required is difficult and time consuming, and agreements wrought can quickly go out of date as new technologies emerge.

What do we gain as humans by risking robotic and AI-initiated disruption and by accommodating the many environmental and social changes needed to implement robots and autonomous vehicles in our lives? As well as exploring the research efforts to human-proof robots, this book has also emphasized human survival and wellbeing in ever more hostile and alien environments. Humans increasingly work and live in spaces that are being redesigned with robots and autonomous vehicles in mind with few thoughts about what it will do to their fragile bodies and minds. As characterized by roboticist Daniel H. Wilson, "Sometimes a technology is so awe-inspiring that the imagination runs away with it - often far, far away from reality. Robots are like that. A lot of big and ultimately unfulfilled promises were made in robotics early on, based on preliminary successes" (quoted in Sofge, 2009). The dreams and insights of roboticists and AI researchers have indeed had considerable impacts, but often at the price of establishing visions that are not buttressed by everyday successes and that are occasionally undermined by colossal failures. Today the discernments and insights of technically-aware individuals are needed in order to shape the technologies in ways that support the autonomy and wellbeing of individuals and the smooth functioning of societies. The hard work of making robotics, autonomous vehicles, and AI suitable for the fulfillment of human needs will occupy a great deal of the century to come.

References

Abnet, D. A. (2020). *The American robot: A cultural history*. University of Chicago Press.

Alexander, S. (2019, March 8). Robots 'will be commonplace in homes by 2050' with 'android rights and PAY.' *Mirror*. https://www.mirror.co.uk/news/weird-news/robots-will-commonplace-homes-2050-14105140

Asimov, I. (1940). *I, Robot*. Narkaling Productions.

Asimov, I. (1986). *Robots and empire*. Ballantine.

Baczko, B. (1989). *Utopian lights*. Paragon House.

Baker, B. D. (2020). Sin and the Hacker Ethic: The tragedy of techno-utopian ideology in cyberspace business cultures. *Journal of Religion and Business Ethics, 4*(2), 1.

Beatty, C. A., & Gordon, J. R. (1991). Preaching the gospel: The evangelists of new technology. *California Management Review, 33*(3), 73–94.

Beckwith, B. (1984). *Ideas about the future: A history of futurism, 1794–1982*. Burnham P. Beckwith Publisher.

Belk, R. (2016a). Understanding the robot: Comments on Goudey and Bonnin (2016a). *Recherche et Applications En Marketing (English Edition) (Sage Publications Inc.), 31*(4), 83–90. https://doi.org/10.1177/2051570716658467

Belk, R. (2016b). Understanding the robot: Comments on Goudey and Bonnin (2016). *Recherche Et Applications En Marketing, 31*(4), 83–90. https://doi.org/10.1177/2051570716658467

Belk, R., Weijo, H., & Kozinets, R. V. (2021). Enchantment and perpetual desire: Theorizing disenchanted enchantment and technology adoption. *Marketing Theory, 21*(1), 25–52.

Bellamy, E. (1888). *Looking backward*. Ticknor.

Birhane, A., & van Dijk, J. (2020, February). Robot rights? Let's talk about human welfare instead. In *Proceedings of the AAAI/ACM Conference on AI, Ethics, and Society* (pp. 207–213).

Bisconti Lucidi, P., & Nardi, D. (2018, December). Companion robots: The hallucinatory danger of human-robot interactions. In *Proceedings of the 2018 AAAI/ACM Conference on AI, Ethics, and Society* (pp. 17–22).

Bolton, M. B., & Mitchell, C. C. (2020). When scientists become activists: The international committee for robot arms control and the politics of killer robots. In *Global activism and humanitarian disarmament* (pp. 27–58). Palgrave Macmillan, Cham.

Bradbury, R. (1969). *I Sing the Body Electric (and other stories)*. Knopf Publishing Co.

Brautigan, R. (1968). All watched over by machines of loving grace. *TriQuarterly, 11*, 194.

Bringsjord, S. (2008). Ethical robots: The future can heed us. *AI & Society, 22*(4), 539–550.

Brodt, S. E., & Tuchinsky, M. (2000). Working together but in opposition: An examination of the "good-cop/bad-cop" negotiating team tactic. *Organizational Behavior and Human Decision Processes, 81*(2), 155–177.

Brown, D. (2021, July 27). Toyota's basketball robot stuns at the Tokyo Olympics with its flick of the wrist. *Washington Post*. https://www.washingtonpost.com/technology/2021/07/27/tokyo-olympics-robot/

Campbell, J., & Moyers, B. (2011). *The power of myth*. Anchor.

Cellan, R. (2014). Stephen Hawking warns artificial intelligence could end mankind. *BBC News, 2*(2014), 10.

Coates, J. F. (2000). Looking ahead: There is no Joy in my life. *Research Technology Management, 43*(6), 6–8.

Dewitte, M., & Reisman, Y. (2021). Clinical use and implications of sexual devices and sexually explicit media. *Nature Reviews Urology, 18*(6), 359–377.

Dick, P. K. (1955). *Spring)*. Parigi Books.

Dinello, D. (2021). *Technophobia!* University of Texas Press. https://doi.org/10.7560/709546-002

Donhauser, J. (2019). Environmental robot virtues and ecological justice. *Journal of Human Rights and the Environment, 10*(2), 176–192.

Duan, H., Li, J., Fan, S., Lin, Z., Wu, X., & Cai, W. (2021, October). Metaverse for social good: A university campus prototype. In *Proceedings of the 29th ACM International Conference on Multimedia* (pp. 153–161).

Gates, B. (2007). A robot in every home. *Scientific American, 296*(1), 58–65.

Gibson, W. (1986). *Neuromancer*. Berkley.

Hamilton, I. A. (2022, February 16). Mark Zuckerberg's new values for Meta show he still hasn't truly let go of 'move fast and break things.' *Business Insider*. https://www.businessinsider.in/tech/news/mark-zuckerbergs-new-values-for-meta-show-he-still-hasnt-truly-let-go-of-move-fast-and-break-things/articleshow/89623917.cms

Hancock, P. A. (2019). Some pitfalls in the promises of automated and autonomous vehicles. *Ergonomics, 62*(4), 479–495.

Haraway, D. (1987). A manifesto for cyborgs: Science, technology, and socialist feminism in the 1980s. *Australian Feminist Studies, 2*(4), 1–42. https://doi.org/10.1080/08164649.1987.9961538

Harris, D. (1994). The aesthetic of the computer. *Salmagundi,* (101/102), 173–181.

Heffernan, T. (2020). The dangers of mystifying artificial intelligence and robotics. *Toronto Journal of Theology, 36*(1), 93–95.

Hencley, S. (1974). *Futurism in education: Methodologies.* Berkeley, CA: McCutchan.

Huxley, A. (1932). *Brave new world.* Chatto & Windus.

Jackson, R. D. (2020). "I approved it... and I'll do it again": Robotic policing and its potential for increasing excessive force. *Proceedings of the ETHICOMP 2020.* ISBN 978-84-09-20272-0. 347–349

Jasanoff, S. (2021). Humility in the anthropocene. *Globalizations, 18*(6), 839–853. https://doi.org/10.1080/14747731.2020.1859743

Joy, B. (2020). Why the future doesn't need us: Our most powerful 21st-century technologies-robotics, genetic engineering, and nanotech-are threatening to make humans an endangered species. In *Emerging Technologies: Ethics, Law and Governance* (pp. 47–63). Routledge.

Kang, M. (2011). *Sublime dreams of living machines.* Harvard University Press.

Kietzmann, J., & Pitt, L. F. (2020). Artificial intelligence and machine learning: What managers need to know. *Business Horizons, 63*(2), 131–133.

Komatsu, T., Malle, B. F., & Scheutz, M. (2021, March). Blaming the reluctant robot: Parallel blame judgments for robots in moral dilemmas across US and Japan. In *Proceedings of the 2021 ACM/IEEE International Conference on Human-Robot Interaction* (pp. 63–72).

Körtner, T. (2016). Ethical challenges in the use of social service robots for elderly people. *Zeitschrift Für Gerontologie Und Geriatrie, 49*(4), 303–307.

Kuznetsov, S., Paulos, E., & Gross, M. D. (2010, August). WallBots: interactive wall-crawling robots in the hands of public artists and political activists. In *Proceedings of the 8th ACM Conference on Designing Interactive Systems* (pp. 208–217).

LaFrance, A. (2016, March). What is a robot? The question is more complicated than it seems. *The Atlantic.* https://www.theatlantic.com/technology/archive/2016/03/what-is-a-human/473166/

Laham, M. (2020). *Made up: How the beauty industry manipulates consumers, preys on women's insecurities, and promotes unattainable beauty standards.* Rowman & Littlefield Publishers.

Lacković, N. (2021). Postdigital living and algorithms of desire. *Postdigital Science Education 3,* 280–282. https://doi.org/10.1007/s42438-020-00141-4

Levy, D. (2009). *Love and sex with robots: The evolution of human-robot relationships*. Harper.

Liu, H. Y. (2018). The power structure of artificial intelligence. *Law, Innovation and Technology, 10*(2), 197–229.

Majeed, A., Baadel, S., & Haq, A. U. (2017, January). Global triumph or exploitation of security and privacy concerns in e-learning systems. In *International Conference on Global Security, Safety, and Sustainability* (pp. 351–363). Springer. https://doi.org/10.1007/978-3-319-51064-4_28

Mainenti, D. (2020). Sex robot technology and the Narrative Policy Framework (NPF): A relationship in the making? *Paladyn, Journal of Behavioral Robotics, 11*(1), 390–403. https://doi.org/10.1515/pjbr-2020-0022

Mamak, K. (2021). Whether to save a robot or a human: On the ethical and legal limits of protections for robots. *Frontiers in Robotics and A, 1*, 8.

Markowitz, J. A. (2019). *Robots that kill: Deadly machines and their precursors in myth, folklore, literature, popular culture and reality*. Jefferson, NC: McFarland.

Massa, F. G., Helms, W. S., Voronov, M., & Wang, L. (2017). Emotions uncorked: Inspiring evangelism for the emerging practice of cool-climate winemaking in Ontario. *Academy of Management Journal, 60*(2), 461–499.

McBride, J. (2021). Climate change, global population growth, and humanoid robots. *Journal of Future Robot Life, 2*(1–2), 23–41.

McDonald, H. (2019, September 15). Ex-Google worker fears 'killer robots' could cause mass atrocities. *The Guardian*. https://www.theguardian.com/technology/2019/sep/15/ex-google-worker-fears-killer-robots-cause-mass-atrocities

Middleton, C. (2021, July 19). The five laws of robotic software automation. *Diginomica*. https://diginomica.com/five-laws-robotic-software-automation

Milligan, L. M. (2013). The forgotten right to be secure. *Hastings Law Journal, 65*(3), 713–760.

Minsky, M. (1994, October). Will robots inherit the Earth? *Scientific American, 271*(4), 109–113. http://web.media.mit.edu/~minsky/papers/sciam.inherit.html

Mitchell, S. (2018). Narratives of resistance and repair in consumer society. *Third Text, 32*(1), 55–67.

Nader, R. (1965). *Unsafe at any speed*. Grossman Publishers.

Nagenborg, M. (2020). Can we forgive a robot? In *Technology, anthropology, and dimensions of responsibility* (pp. 147–157). JB Metzler.

Nees, M. A. (2019). Safer than the average human driver (who is less safe than me)? Examining a popular safety benchmark for self-driving cars. *Journal of Safety Research, 69*, 61–68.

Newton, D. P., & Newton, L. D. (2019, November). Humanoid robots as teachers and a proposed code of practice. In *Frontiers in education* (Vol. 4, p. 125). Frontiers.

"NICE sets the standard" (2021, June 30). *BusinessWire*, https://www.businesswire.com/news/home/20210630005419/en/NICE-Sets-the-Standard-for-Responsible-Design-and-Deployment-of-AI-Powered-Robots-by-Unveiling-its-Robo-Ethical-Framework

Nomura, T., Kanda, T., & Yamada, S. (2019, March). Measurement of moral concern for robots. In *2019 14th ACM/IEEE International Conference on Human-Robot Interaction (HRI)* (pp. 540–541). IEEE Press.

Nourbakhsh, I. R. (2015). The coming robot dystopia. *Foreign Affairs, 94*(4), 23–28.

NSPE (2022). *Professional policies and position statements.* https://www.nspe.org/resources/issues-and-advocacy/professional-policies-and-position-statements/autonomous-vehicles

Olaronke, I., Rhoda, I., Gambo, I., Oluwaseun, O., & Janet, O. (2020). A systematic review of swarm robots. *Current Journal of Applied Science and Technology, 39*, 79–97.

Oravec, J. A. (1996). *Virtual individuals, virtual groups: Human dimensions of groupware and computer networking.* Cambridge University Press.

Oravec, J. A. (1998). Futurism in the organizational and end-user information systems curriculum: Critical perspectives and classroom exercises, *OSRA Journal, 16*(2), 35–42 (EJ576761).

Oravec, J. A. (2018). Cyberloafing and constructive recreation. In *Encyclopedia of information science and technology, fourth edition* (pp. 4316–4325). Hershey, PA: IGI Global.

Oravec, J. A. (2021). Robots as the artificial "other" in the workplace: Death by robot and anti-robot backlash. *Change Management: An International Journal, 21*(2). https://doi.org/10.18848/2327-798X/CGP/v21i02/65-

Oravec, J. A. (2022b, forthcoming). AI, biometric analysis, and emerging cheating detection systems: The engineering of academic integrity? *Educational Policy Analysis Archives, 27*, (pp. 8413–8460).

Oravec, J. A. (2014). Mottos and ethical statements of Internet-based organizations: Implications for corporate social responsibility. *International Journal of Civic Engagement and Social Change (IJCESC), 1*(2), 37–53.

Oravec, J. A. (2017). Kill switches, remote deletion, and intelligent agents: Framing everyday household cybersecurity in the Internet of Things. *Technology in Society, 51*, 189–198.

Oravec, J. A. (2019). Artificial intelligence, automation, and social welfare: Some ethical and historical perspectives on technological overstatement and hyperbole. *Ethics and Social Welfare, 13*(1), 18–32.

Oravec, J. A. (2022a). The emergence of "truth machines"?: Artificial intelligence approaches to lie detection. *Ethics and Information Technology, 24*(1), 1–10. https://doi.org/10.1007/s10676-022-09621-6

Orwell, G. (1948). *1984*. Harcourt.

Paiva, A., Leite, I., Boukricha, H., & Wachsmuth, I. (2017). Empathy in virtual agents and robots: A survey. *ACM Transactions on Interactive Intelligent Systems (TiiS), 7*(3), 1–40.

Parrinder, P. (2009). Robots, clones and clockwork men: The post-human perplex in early twentieth-century literature and science. *Interdisciplinary Science Reviews, 34*(1), 56–67.

Rainie, L., Anderson, A., & Vogels, E. (2021, June 16). *Experts doubt ethical AI design will be broadly adopted as the norm within the next decade*. Pew Research Center. https://www.pewresearch.org/internet/2021/06/16/experts-doubt-ethical-ai-design-will-be-broadly-adopted-as-the-norm-within-the-next-decade/

Ramana, M. V. (2018). Technical and social problems of nuclear waste. *Wiley Interdisciplinary Reviews: Energy and Environment, 7*(4). https://doi.org/10.1002/wene.289

Reich-, N., Eyssel, F., & Hohnemann, C. (2019). Involve the user! Changing attitudes toward robots by user participation in a robot prototyping process. *Computers in Human Behavior, 91*, 290–296.

Reilly, K. (2019). Robots and anthropomorphism in science-fiction theatre: From rebellion to domesticity and back again. In *Media archaeology and intermedial performance* (pp. 193–210). Palgrave Macmillan, Cham.

Rodger, I. (1965). The headless horseman: An amateur inquiry. *Journal of the Folklore Institute, 2*(3), 266–271.

Rose, B. A. (1996). *Jekyll and Hyde adapted: Dramatizations of cultural anxiety* (No. 66). Greenwood Publishing Group.

Ruggiu, D. (2018). Implementing a responsible, research and innovation framework for human enhancement according to human rights: The right to bodily integrity and the rise of 'enhanced societies.' *Law, Innovation and Technology, 10*(1), 82–121.

Salzmann-Erikson, M., & Eriksson, H. (2016). Tech-resistance: The complexity of implementing nursing robots in healthcare workplaces. *Contemporary Nurse: A Journal for the Australian Nursing Profession, 52*(5), 567.

Sanoubari, E., Muñoz Cardona, J. E., Mahdi, H., Young, J. E., Houston, A., & Dautenhahn, K. (2021, June). Robots, bullies and stories: A remote co-design study with children. In *Interaction design and children* (pp. 171–182).

Santosh, K. C., & Gaur, L. (2021). Privacy, security, and ethical issues. In *Artificial intelligence and machine learning in public healthcare* (pp. 65–74). Springer.

Santow, E. (2019). *Human rights and technology: Discussion paper*. Australian Human Rights Commission. https://humanrights.gov.au/our-work/rights-and-freedoms/publications/human-rights-and-technology-discussion-paper-2019

Scharrer, L., Rupieper, Y., Stadtler, M., & Bromme, R. (2017). When science becomes too easy: Science popularization inclines laypeople to underrate their dependence on experts. *Public Understanding of Science, 26*(8), 1003–1018.

Schieber, P. (1987, March/April). The wit and wisdom of Grace Hopper. *The OCLC Newsletter, 167*. http://www.cs.yale.edu/homes/tap/Files/hopper-wit.html

Schizer, M. (2021, November 2). Henry Kissinger says AI is "as consequential" but less predictable than nuclear weapons. *Newsweek*. https://www.newsweek.com/2021/11/12/henry-kissinger-says-ai-consequential-lesspredictable-nuclear-weapons-1644508.html

Schuster, J., & Woods, D. (2021). *Calamity theory: Three critiques of existential risk*. University of Minnesota Press.

Seligman, M. E. (1972). Learned helplessness. *Annual Review of Medicine, 23*(1), 407–412.

Sharma, B. (2021, May 30). First instance of 'killer robot' attacking human without orders recorded in Libya: UN report. *Wionews*. https://www.wionews.com/technology/first-instance-of-killer-robot-attacking-human-without-orders-recorded-in-libya-un-report-388344

Sheinkopf, K. G., & Weintz, M. R. (1973). The Beatles are dead! Long live the Beatles! *Popular Music & Society, 2*(4), 321–326.

Simonite, T. (2021, June 3). Don't end up on this artificial intelligence hall of shame. *Wired*, https://www.wired.com/story/artificial-intelligence-hall-shame/

Smids, J., Nyholm, S., & Berkers, H. (2020). Robots in the workplace: A threat to—or opportunity for—meaningful work? *Philosophy & Technology, 33*(3), 503–522.

Sofge, E. (2009, October 1). Robotics expert Q&A: Daniel H. Wilson. *Popular Mechanics*. https://www.popularmechanics.com/technology/robots/a1224/4210553/

Valente, C. M. (2012). *Silently and very fast*. Cape Town Books.

van Pinxteren, M. M., Wetzels, R. W., Rüger, J., Pluymaekers, M., & Wetzels, M. (2019). Trust in humanoid robots: implications for services marketing. *Journal of Services Marketing*. https://www.emerald.com/insight/content/doi/https://doi.org/10.1108/JSM-01-2018-0045/full/html

Vanderelst, D., & Winfield, A. (2018, December). The dark side of ethical robots. In *Proceedings of the 2018 AAAI/ACM Conference on AI, Ethics, and Society* (pp. 317–322). https://doi.org/10.1145/3278721.3278726

Vichitkraivin, P., & Naenna, T. (2021). Factors of healthcare robot adoption by medical staff in Thai government hospitals. *Health and Technology, 11*(1), 139–151.

Wang, W., & Siau, K. (2019). Artificial intelligence, machine learning, automation, robotics, future of work and future of humanity: A review and research agenda. *Journal of Database Management (JDM), 30*(1), 61–79.

Webb, S. (2017). Mad scientists. In *All the Wonder that Would Be* (pp. 297–315). Springer, Cham.

Wells, H. (1936). *The shape of things to come.* Macmillan.

Whitby, B. (2008). Sometimes it's hard to be a robot: A call for action on the ethics of abusing artificial agents. *Interacting with Computers, 20*(3), 326–333.

Wike, R., & Stokes, B. (2018, September 13). In advanced and emerging economies alike, worries about job automation. *Pew research center.* https://www.pewresearch.org/global/2018/09/13/in-advanced-and-emerging-economies-alike-worries-about-job-automation/

Zeller, B., Trakman, L., & Walters, R. (2019). The Internet of Things-The Internet of Things or of human objects: Mechanizing the new social order. *Rutgers Law Record, 47,* 15–103.

Index

A
Abundance economics, 167
Addiction, 19, 72, 74, 103, 109
Advanced Research Projects Agency, US (ARPA), 157
AI incident database. *See* Artificial Intelligence (AI)
AI winter. *See* Artificial Intelligence (AI)
Algorithm, 26, 51, 99, 162, 168, 250, 257, 263
Alphabet Corporation (US)
  DeepMind, 162, 163
  Google, 39, 57, 70, 158, 162, 163, 166, 167, 195, 268
  LaMDA (computer system), 39
Amazon Corporation (US), 221
Ameca (robot), 15
American Civil Liberties Union (ACLU), 130
Anthropomorphism, 210, 221, 227, 267

Anxiety, 3, 11, 13, 16, 18, 22, 28, 40, 70, 106, 125, 132, 133, 138, 186, 212, 254
Apple (US corporation), 158, 167
Armageddon, AI. *See* Artificial Intelligence (AI)
Army, US, 126
Artificial Intelligence (AI)
  AI Incident Database, 182
  AI Winter, 69, 164
  Armageddon, 52, 186
  expert systems, 8, 160, 161, 164, 165
  facial recognition, 125, 248
  hyperbole, 40, 153–155, 165, 167, 168, 191
  knowledge engineering, 160, 164
  machine learning, 8, 26, 141, 158, 161, 182, 189, 215
Asimov, Isaac
  I, Robot, 50
  The Robots of Dawn, 99
  Three Laws of Robotics, 184, 215

Australian Human Rights
  Commission, 262
Automation, 41, 44, 63, 91, 153,
  156, 162, 167, 179–181, 183,
  190, 192, 193, 206–208, 213,
  215, 218, 220, 223–226, 255
Autonomous systems, 61, 126, 128,
  159, 160, 163, 178–180, 191,
  192, 194
Autonomous vehicles, 2, 3, 5, 6,
  11–13, 18–20, 22, 24, 27, 30,
  40, 45, 48, 49, 54, 62, 69,
  71–74, 94, 100, 115, 125, 131,
  133, 136, 141, 153, 161, 167,
  178, 179, 185, 186, 194, 205,
  207–211, 213, 214, 217, 219,
  224–226, 245–250, 254, 255,
  257–269
  self-driving cars, 27, 46, 71, 178,
    179, 210

B

Barlow, John Perry, 146
BattleBots, 213, 224
Bellamy, Edward, 249
Bias, 6, 67, 167, 216, 218, 251, 255,
  258, 263
Bicentennial Man, 99
Big data, 59, 166
Biometrics, 102, 108, 114, 125
Black Mirror, 25
Boston Dynamics (US corporation),
  40, 47, 207
Bourdieu, Pierre, 212
Bradbury, Ray, 228, 265
Brautigan, Richard, 253
Bullying, 4, 205, 207, 209, 218, 219,
  221, 226, 227

C

Campaign Against Sex Robots
  (CASR), 24, 105
Campaign to Stop Killer Robots, 24,
  137
Campbell, Joseph, 245
Čapek, Karel, 13, 14, 183, 215
Carnegie Mellon University, 26, 45
Centre for the Study of Existential
  Risk (at the University of
  Cambridge, UK), 261
Chatbots, 44, 94, 96, 97, 109, 110,
  158, 187, 209, 218, 220
China, 30, 127, 145
Clarke, Arthur C., 56, 184, 215, 228
Coates, J.F., 249
Commodification, 91, 107–109
Compulsory robot usage, 41, 125
Computer Professionals for Social
  Responsibility (CPSR), 7, 8, 139,
  143–146, 260
Contests, 5, 12, 49, 51, 52, 62, 63,
  72, 111, 159, 161, 168, 169,
  188, 192, 224, 246, 266
Control, 3, 5, 12, 14, 18, 20, 24, 27,
  28, 44–46, 48, 60, 62, 67, 70,
  72, 91, 92, 95, 98, 102,
  107–109, 112, 114–116, 125,
  127, 129, 131, 132, 135–137,
  139–142, 158, 160, 163, 178,
  183, 184, 189, 192, 211, 213,
  215, 216, 218, 223, 246, 255,
  259, 263, 264, 266
COVID-19 pandemic. *See* Pandemic
Craigslist, 96
Creepiness, 5, 18, 19, 27, 39, 58–61,
  63, 71, 92, 113, 114, 178
Cybernetics, 45, 126, 158, 223
Cybersecurity. *See* Security
Cyborg, 6, 25, 45, 53, 99, 145, 216,
  223, 227, 256

D

Dark patterns, 2, 16, 17, 24, 42, 53, 54, 62, 93, 94, 114, 247, 251, 253
da Vinci, Leonardo, 5
Death by robot, 177, 179
DeepMind. *See* Alphabet Corporation (US)
Defense Advanced Research Projects Agency, US (DARPA), 157
Dehumanization, 25, 226
Delvaux-Stehres, Mady, 168
Deskilling, 131, 136, 167
Dhake, Umesh Ramesh, 183, 194
Dick, Philip K., 265
Diebold, John, 215
Disney (US corporation), 19
Dissecting a robot, 25, 26
Dogs (as robots), 1, 54, 130, 136
Domestic robots. *See* Robots
"Don't be evil", 167, 268
Douglas, William O., 140
Dramaturgical
   frontstage and backstage, 15, 112
   humor as catharsis, 103
   leakage, 16
Dreyfus, Hubert, 166
Drones, 131, 135, 141, 156, 180, 184, 205, 208, 211
Dystopian scenarios, 249, 252

E

Electronic Frontier Foundation- (EFF), 146, 260
Emerson, Ralph Waldo, 43
Engelbart, Douglas, 68
Engelberger, Joseph, 67
Engineered Arts (UK corporation), 15
Erobotics, 17, 92
Ethics
   codes of ethics, 144, 167
   moral imperative, 166
   moral panic, 92, 104
   robot rights, 29
European Court of Human Rights, 248
Evil, 5, 25, 42, 55, 56, 58, 64, 66, 114, 195
EX Doll (robot), 26, 104
Ex Machina, 105
Expert systems. *See* Artificial intelligence (AI)

F

Facebook. *See* Meta
Facial recognition. *See* Artificial Intelligence (AI)
Fear amplification, 132, 133
Feigenbaum, Edward, 160, 164
Ford Motor Company (US corporation), 139, 180
Forsythe, Diana, 65, 154, 158
Fourth Amendment. *See* US Constitution
Frankenstein, 53, 65, 184, 216
Freedom of association, 43
Fromm, Erich, 4
Frubber, 26
Funerals (for robots), 179, 187, 214, 223
Future of Humanity Institute (University of Oxford, UK), 268
The Future of Life Institute (FLI), 126, 138
Future shock, 13, 23
Futurism, 23

G

Gaslighting, 94, 95, 218
Gates, Bill, 168, 250

Gender, 5, 28, 44, 67, 74, 99, 101, 103, 106, 108, 187, 218, 247, 255, 258, 263
Gibson, William, 250
Goffman, Erving, 16, 103, 184, 226
Google. *See* Alphabet
GPT-3 (robot), 39
Guarded trust. *See* Trust

H
HAL (robot), 20
Hanson Robotics (Hong Kong company), 39, 41
Haraway, Donna, 6, 45, 53
Hate crimes, 29, 207, 211
Herzberg, Elaine, 24, 179, 194
hitchBOT (robot), 209
Holbrook, Wanda, 24, 177, 194
Hopper, Grace Murray, 248
Huxley, Aldous, 249
Hyperbole. *See* Artificial Intelligence (AI)
Hyundai (South Korean multinational manufacturer), 17, 40, 47

I
IBM (US corporation), 50, 51, 62, 158, 159, 167, 188
  Deep Blue, computer system, 51
  Watson, computer system, 50, 62, 159, 188
I'm Your Man, 100
Individual rights. *See* Ethics
Industrialization, 13
Infantilization, 23
International Committee of the Red Cross (ICRC), 138
International Congress on Love and Sex with Robots, 105
International Federation of Robots (IFR), 40, 44, 183, 215

Internet of Things (IoT), 44, 218
Internet of Robotic Things (IoRT), 23, 44, 45
I, Robot. *See* Asimov, Isaac

J
Japan, 69, 93, 180, 207, 265
Jibo (robot), 70

K
Kapor, Mitch, 146
Kasparov, Garry, 51
K5 (robot), 209
Killer robots. *See* Robots
Kissinger, Henry, 134, 249
Knightscope security robots, 209
Knowledge engineering. *See* Artificial Intelligence (AI)
Kurzweil, Ray, 57, 258

L
Lal, Ramji, 180
LaMDA (computer system). *See* Google
Lang, Fritz, 14, 25
Lanier, Jaron, 4
Leakage. *See* Dramaturgical
Learned helplessness, 262
Lethal Autonomous Weapon Systems (LAWS), 126, 138, 186
Levy, David, 17, 49, 92, 96
Li, Fei-Fei, 156
Lighthill Report (UK), 164
Lovelace, Ada, 169
Lovotics, 17, 92
Ludd, Ned, 208

M
Machine learning. *See* Artificial Intelligence (AI)

Machine wrecking, 208, 214
Magic, 56, 189
 magical thinking, 252
Magical thinking. *See* Magic
Mark of the beast, 56
Marriage (human-robot), 17, 93, 98–101, 250
Marvel Cinematic Universe, 27
McCarthy, John, 156, 157
McLuhan, Marshall, 43, 145
Meta (US corporation), 26, 40, 65, 268
Metaverse, 6, 13, 26, 40, 47, 113, 225, 246, 255, 268
Metropolis, 14, 25
Microsoft (US corporation), 168, 250
Military robots. *See* Robots
Mills, C. Wright, 29
Minsky, Marvin, 53, 69, 155, 245
Moral imperative. *See* Ethics
Moral panic. *See* Ethics
Mori, Masahiro, 59, 193, 216, 221
Musk, Elon, 11, 68, 222, 268

N
Nader, Ralph, 258
 Unsafe at Any Speed, 258
National Institute for Occupational Safety and Health, US (NIOSH), 181, 182
National Society of Professional Engineers, 261
Newell, Allen, 156, 160
NICE (Israeli corporation), 263, 264
1984, 249
Norman, Donald, 59, 60, 74

O
OnlyFans, 96
OpenAI, 39
Orphaning (of technologies), 69, 71

Orwell, George, 249
"The other", 4, 13, 46, 59, 65, 189
Overtrust. *See* Trust

P
Pandemic, 41, 135, 138
 COVID-19 pandemic, 96
 Covid pandemic, 24
Pepper (robot), 40, 70
Police robot. *See* Robots
Pope Francis, 58, 138
Pornography, 107, 109
Post-human, 53
Privacy, 68, 92, 93, 113, 140, 144, 145, 167, 193, 258
Projection (psychological), 13, 21
Psychopathy, 59
Pygmalion, 101, 110, 112, 115

Q
Quantified self, 112
Quartering (of soldiers), 139, 140, 246

R
Ray, Amit, 17, 58, 98
Ready Player One, 47
RealDoll Corporation, 93
Religion, 58, 247, 263
Resistance, 24, 125, 141, 223, 224, 259, 264, 266
ReSkin, 26
Responsibility, 19, 62, 142, 161, 194, 224, 260, 268
Richardson, Kathleen, 21, 105, 106, 108, 113, 156, 181, 214, 221
Road rage, 207, 211
Robo-hype, 153
Robophobia, 27, 184
Robosexual, 92

Robot Hall of Fame, 45
Robotland (South Korea), 47
Robot rights. *See* Ethics
Robots
   domestic robots (dombots), 72, 219, 220, 247
   killer robots, 6, 24, 28, 43, 51, 53, 54, 60, 66, 126, 127, 130, 132, 133, 137, 138, 140–142, 186, 259, 261, 267
   manufacturing robots, 44, 94, 114, 115, 183, 215, 220
   military robots, 21, 55, 64, 72, 74, 127, 131, 186, 195, 246, 266, 267
   police robots, 8, 24, 60, 72, 92, 95, 116, 127, 130, 136, 141–143, 180, 261
   rogue robots, 14, 53, 189, 252, 267
   service robots, 3, 4, 40, 44, 49, 61, 94, 114, 115, 140, 183, 187, 192, 210
   sex robots, 1, 17, 19–21, 24, 26, 28, 55, 61, 64, 72, 74, 91–116, 225, 247, 255, 267
   social robotics, 61–63, 73, 106, 108
   swarm robots, 61, 248
Robot taxes, 166, 167
Rogue robots. *See* Robots
Roll-Oh (robot), 14
Rosenblatt, Frank, 158
Rossum's Universal Robots (RUR), 13, 14, 28, 183, 215
Roxxxy (robot), 96, 97
Rumination time, 22

S
Sabotage, 22, 40, 179, 194, 205–208, 210, 211, 223–227
Saudi Arabia, 40

Schuler, Douglas, 8
Scientific Management. *See* Taylorism
Scientist, celebrity, 68
Security, 7, 18, 23, 27, 48, 55, 56, 61, 68, 71, 91, 94, 115, 125–127, 129–131, 133, 135, 136, 139, 140, 142–144, 169, 178, 184, 189, 190, 194, 195, 205, 209, 210, 214, 220, 222, 227, 246, 247, 251, 254, 255, 257, 260, 262
   cybersecurity, 48, 251
Self-driving cars. *See* Autonomous vehicles
Sex robots. *See* Robots
Sex workers, 96, 104, 107, 212
Shelley, Mary, 18, 53, 65, 184, 216
Silicon Valley, 48
Simon, Herbert, 155, 157, 160
Singapore, 131
Singularity, 46, 50, 57, 58, 93, 98, 99, 252
Slaughterbots, 126
Slavery, 21, 205, 219
Social engineering, 74, 95, 267
Social protest, 212
Social Robotics. *See* Robotics
SoftBank (Japanese multinational conglomerate), 40, 208
Sophia (robot), 39, 40
Space Force (US), 126
Spiritual, 17, 57, 58, 93, 98, 99, 101, 252–254
Stephenson, Neal, 6, 47
Strategic Defense Initiative (SDI), 145, 154
Supernatural, 46, 55, 57, 58
Surveillance, 5, 42, 46, 56, 59, 91, 136, 140, 181, 191, 220, 264
Swarm robots. *See* Robots
Symbolic violence, 206, 212, 213, 227, 228

T
Taylorism, 164
  Scientific Management, 164
Technological visions, 168, 191
Technosex, 92, 95–98, 113–115
The Terminator, 14, 52
Terrorism, 192
Tesla (US corporation), 11, 43, 68
Texas Instruments (US corporation), 165
Thinking machine, 158, 159, 162, 169, 187, 188, 216, 217
Third Amendment. *See* US Constitution
Three Laws of Robotics. *See* Asimov, Isaac
Toffler, Alvin, 23
TrueCompanion (US corporation), 96, 97
Trust, 23, 46, 184, 189–192, 194, 217, 219, 254, 257
  guarded trust, 193
  overtrust, 16
Turing, Alan, 50, 157
2001: A Space Odyssey, 20, 184, 216

U
Ultron (robot), 27

Uncanny valley, 58, 59, 193, 215, 221
United Nations (UN), 53, 137
Unsafe at Any Speed. *See* Nader, Ralph
Urada, Kenji, 180
US Constitution, 139
  Third Amendment, 140
  Fourth Amendment, 139
Utopian scenarios, 251

W
Walmart (US corporation), 70
Wareham, Mary, 137
Watson, computer system. *See* IBM (US corporation)
Wearables, health, 93, 96, 102, 112
Weizenbaum, Joseph, 166
Wells, H.G., 249
Westinghouse Electric (US corporation), 14, 25
Wiener, Norbert, 155, 158–160, 170
Williams, Jody, 137
Williams, Robert, 24, 180, 185, 194
Winner, Langdon, 142

Z
Zuckerberg, Mark, 65, 178, 268